T0220467

THE REALISTIC EMPIRICISM OF MACH, JAMES, AND RUSSELL

In the early twentieth century, Ernst Mach, William James, and Bertrand Russell founded a philosophical and scientific movement known as "neutral monism," based on the view that minds and physical objects are constructed out of elements or events which are neither mental nor physical, but neutral between the two. This movement offers a unified scientific outlook which includes sensations in human experience and events in the world of physics under one roof. In this book Erik C. Banks discusses this important movement as a whole for the first time. He explores the ways in which the three philosophers can be connected, and applies their ideas to contemporary problems in the philosophy of mind and the philosophy of science – in particular the relation of sensations to brain processes, and the problem of constructing extended bodies in space and time from particular events and causal relations.

ERIK C. BANKS is Associate Professor of Philosophy in the Department of Philosophy at Wright State University, Ohio. He is also author of *Ernst Mach's World Elements* (2003).

THE REALISTIC EMPIRICISM OF MACH, JAMES, AND RUSSELL

Neutral monism reconceived

ERIK C. BANKS

Wright State University

CAMBRIDGE
UNIVERSITY PRESS

CAMBRIDGE
UNIVERSITY PRESS

University Printing House, Cambridge CB2 8BS, United Kingdom

Cambridge University Press is part of the University of Cambridge.

It furthers the University's mission by disseminating knowledge in the pursuit of education, learning and research at the highest international levels of excellence.

www.cambridge.org
Information on this title: www.cambridge.org/9781107423763

First published 2014
First paperback edition 2016

A catalogue record for this publication is available from the British Library

Library of Congress Cataloguing in Publication data
Banks, Erik C.
The realistic empiricism of Mach, James, and Russell / Erik C. Banks.
pages cm
Includes bibliographical references.
ISBN 978-1-107-07386-9 (Hardback)
1. Empiricism.   2. Mach, Ernst, 1838–1916.   3. James, William, 1842–1910.   4. Russell, Bertrand, 1872–1970.   I. Title.
B816.B36 2014
146'.44–dc23
2014010874

ISBN 978-1-107-07386-9 Hardback
ISBN 978-1-107-42376-3 Paperback

Contents

List of figures

Preface and acknowledgments

This book is an historico-critical look at a realistic form of empiricism which one finds in the philosophy and science of Ernst Mach, William James, and Bertrand Russell. Some of the research in these chapters has appeared in my 2003 book *Ernst Mach's World Elements*, my 2005 article "Kant, Herbart and Riemann" in *Kant-Studien*, my 2010 article "Neutral Monism Reconsidered" in *Philosophical Psychology*, and my 2013 articles "Extension and Measurement: A Constructivist Program from Leibniz to Grassmann" in *Studies in the History and Philosophy of Science A* and "William James' Direct Realism: A Reconstruction" in *History of Philosophy Quarterly*.

Work on this book began during a Fulbright year in 2004–2005 in Germany, when I was a guest at the Max Planck Institute for the History of Science in Berlin. The first drafts of Chapter 6 were written there and I gave a talk which outlined my construction of extended quantities. After I started thinking about William James again, I ran into Michael Levin, waiting for tickets to Shakespeare in the Park, and we discussed my Jamesian version of direct realism in epistemology. His criticism, delivered on the spot, made me think harder about those ideas. In summer 2011, Jirí Wackermann invited me to work on the book during a stay at the Psychophysics Department of the Institute for Frontier Areas in Psychology, Freiburg im Breisgau, and Jirí, Harald Atmansprecher, and Roemer Hartmann gave me feedback on my presentation "Enhanced Physicalism," which I gave at the Institute's Theory Colloquium. Jirí's writings on Mach from the perspective of a working psychophysicist have deeply influenced my understanding of these issues. In 2012, Jordi Cat and Amit Hagar gave me the chance to present the talk "Extended Magnitudes" at the Indiana University Department of History and Philosophy of Science colloquium, where they provided strong feedback and critical comments. I gave another talk on "The Problem of Extension" at the 2012 meeting of the History of Philosophy of Science (HOPOS) society in Halifax and I benefited from those comments in the preparation of the final chapter. Anonymous reviewers for the publisher, and the journals named above, helped me improve the manuscript. Finally, thanks, as always, to my mother Laurene Buckley, art historian, for teaching me how to do research, and for those long ago book sales at the New Fairfield Public Library and at the Strand when I was young, and to my father Richard T. Banks, architect, who taught me how to teach. I will always be grateful for their confidence in me.

I am a naturalist philosopher; for me, philosophy without science is empty and science without philosophy is blind. I see philosophy and science as continuous: methodological and speculative ideas that originate in philosophy, over many years, are gradually refined until they can be articulated quantitatively and tested and so become part of empirical science, at which point their origins are usually forgotten. We'll see several examples in this book. Most of my ideas and source materials come from rooting around in the history of science and following my interests there. Consequently, I think, those who will get the most out of the book are historico-philosophically minded scientists and naturalist philosophers who look to the history of science for their source material.

Conceptually, this book is about "what happens when something happens." It is about *events* in the natural world: events in physical science and events in the brain, which are gathered together under the common term of "elements." I use the historico-critical method to investigate the elements and their relation to science and philosophy. This method, which I learned from Mach's books, broadens and enriches the spectrum of contemporary philosophical ideas and vocabulary, while reconstructing concepts in a rigorous way. I have tried to keep the historical and conceptual goals separate as much as possible in the exposition, but I do hold that my realistic empiricist view is a direct descendent of its historical ancestor in the three original authors of the tradition. The reader can judge if I have been successful. I think the job of philosophy should be to look at the *whole* historical-conceptual spectrum of ideas and ask: what are the possibilities? I hope to show that realistic empiricism is not only an historically significant view, it is also well worth reconsidering today.

Introduction: An overview of realistic empiricism

Introduction

Realistic empiricism is a view of science and philosophy that got its start in the work of Ernst Mach, William James, and Bertrand Russell. The view is also known as neutral monism, radical empiricism, or empirio-criticism. These three authors, for all their differences, share a common historical and conceptual framework, which justifies treating their views as part of the same movement, as I will show in the first three chapters of the book. In the second part, I will redevelop the view for application to contemporary problems. In the present introduction, I will characterize the main ideas of realistic empiricism before entering onto the detailed history of its development and my own additions.

The original authors: Mach, James, and Russell

Ernst Mach (1838–1916) was an Austrian physicist and philosopher who worked in Graz, Prague, and Vienna during the late nineteenth and early twentieth centuries. He is known in physics for his critique of Newtonian absolute space, the principle of inertia, and the mechanical world view, critiques which influenced Einstein, Heisenberg, and many other scientists of the following generation. These criticisms can be found sharply articulated in his *History and Root of the Principle of the Conservation of Energy* (1872/1910) and his *The Science of Mechanics* (1883/1960). As is well known, Mach was skeptical of atoms for most of his career and was one of the last scientists to convert to the atomic theory around 1903 (Blackmore 1992, Banks 2003, pp. 12–14).

Mach was also a pioneer in the area of psychophysics, the scientific measurement of sensations, presenting his results in the *Analysis of Sensations* in 1886. It appears that this double professional competence in physics and psychology pushed him to develop a monistic, umbrella framework for the natural sciences capable of handling human experience and physics under one roof. In accord with this view, Mach stripped down empirical theories into what he called "elements," ordered in spare functions or causal–functional connections of various sorts. Within this framework, sensations were taken to be as real as physical events involving bodies and forces. Others, such as the English mathematician W. K. Clifford and the philosopher Richard Avenarius, were developing similar monist views at the time (see Banks 2003, ch. 9).

In philosophy, Mach is best known for influencing the logical positivists Philipp Frank, Moritz Schlick, Rudolf Carnap, Otto Neurath, and Hans Reichenbach. As I have argued elsewhere (Banks 2003, pp. 9–16 and 2013a) I feel this link is tenuous because Mach's views were much more realistic than is commonly realized. In addition, the logical positivists were pre-occupied with second-order questions about logico-linguistic frameworks, analytic versus synthetic truths, and the role of a priori knowledge, which only began to surface after the rise to prominence of modern logic, questions that are alien, and in some ways even antithetical, to Mach's realistic first-order project to reform physics, as Paul Feyerabend pointed out (1970, 1984).

William James (1842–1910) was an American psychologist and philoso-pher, the author of the famous textbook *The Principles of Psychology* and one of the founding American pragmatists. James and Mach knew each other through their common interest in the sensations of movement and they met personally in 1883 in Prague to exchange ideas. Around the turn of the century, James outlined a view he called "radical empiricism" that owes a great deal to Mach's *Analysis of Sensations*. James also worked out his own direct realist theory of perception, which we will examine in this book, and which turns out, I believe, to have much in common with the empirical realism of Kant (see Banks 2013b).

Bertrand Russell (1872–1970) was the great English philosopher and logician who, with Alfred North Whitehead, placed logic on a firm basis with *Principia Mathematica* and who, with G. E. Moore and Gottlob Frege, founded what we now call analytic philosophy. After his work in "On Denoting" and *Principia*, Russell continued to develop, turning increasingly to questions in the foundations of science and psychology. Russell knew the views of Mach and James very well, and referred to them under the name of "neutral monism," a term he coined in a series of articles for the *Monist* in 1913. Russell began as a skeptic but later converted to the view himself in an essay called "On Propositions" in 1919. In Russell's subsequent works, the *Analysis of Mind* (1921) and the

Analysis of Matter (1927), he further developed neutral monism in a realistic direction, adopting what he called "event particulars" as the basic happenings in nature, encompassing both sensations and physical events. Russell's neutral monism is perhaps the best-known development of what I call the realistic empiricist view, although it is often misread as a form of phenomenalism; and it is often assumed that Russell only held the view for a short time, from around 1919 to 1927, even though Russell strongly denied this (Eames 1967), and recent scholarship bears him out (in particular Lockwood 1981 and Tully 1993). This language about a neutral monist 'movement' inaugurated by Mach stems from Russell himself (Russell 1921, p. 16).

As we shall see, each thinker contributed something unique to the movement, which we miss by treating them in isolation. A unified treatment can show where the movement was headed, and a contemporary update of the position can show how it is useful today. There are, I believe, two important problems in contemporary philosophy where realistic empiricism is relevant: the question of how human sensations relate to physical processes in the brain, and the question of how to construct extended objects and regions from elementary events in the philosophy of physics. These are the subjects of Chapters 5 and 6.

Relation to traditional empiricism and logical positivism

Empiricism is often parodied as the view that "If you can't observe it, it doesn't exist," or that entities not observable in principle are to be expunged from science. The focus is on the sensations of a human observer, or on interactions between an object and a measuring device. This sort of classroom empiricism can be a useful fiction to motivate debate about what is really observed in a certain experiment. These questions can then be used as an entering wedge to introduce more fundamental ideas. In truth, however, empiricism covers a very broad spectrum of views, from the nominalism of William of Ockham to British empiricism, to logical positivism, to W. V. Quine and Bas van Fraassen. Each incarnation is worthy of study, and, as recent scholarship demonstrates, there are often surprises behind the historical stereotypes (on logical positivism, see especially Stadler 1997/2001, Friedman 1999, Uebel 2007, Richardson and Uebel 2007, Banks 2013a). I will not seek to compare and contrast all of these varieties of empiricism in this book, or try to define empiricism in general. Nor will I deal with every author and movement that might be brought into consideration, such as Clifford, Avenarius, the

American Realists, and other figures. I will characterize only what I call the realistic empiricist movement from Mach to Russell (from about 1872 to 1927) and distinguish it from other views only when I think significant differences with traditional empiricism should be pointed out. Some of these differences are so great, however, that they should be stated immediately:

1. Traditional empiricism emphasizes a class of observed events and privileges these events over others. Realistic empiricism gradually broadens the area of interest to include all events or interactions in nature so long as they are causally continuous with observation. But it does not insist on a fundamental distinction between observed and unobserved events. Nor does it insist on a fundamental order of experienced givens, from which to logically construct the unobserved external world. The neutral elements of realistic empiricism are real natural events expressing dynamical force in causal relations to each other, not passive sense data, or loosely associated Humean impressions. Realistic empiricism thus broadens the traditional empiricist category of object and observer interaction, and extends it to the rest of nature, while keeping the continuous causal link to human observation which is empiricism's great strength.

2. Realistic empiricism is *not* a second-order study of the methods, language, or structure of what "science says." It does not offer models of explanation, canons of methodology, or a rational reconstruction of theories and methods in use in science. Instead, we get a first-order theory of real events and causal–functional connections, an ontology of the world that is intended to frame a program for designing specific empiricist theories in science that can then be tested.

3. Realistic empiricism is a theory of the empirical content of a science, based on its austere element-and-function ontology. It is not a theory of a priori structural or linguistic frameworks. The view has little to say about the abstract conceptual framework of science, which is unquestionably part of the enterprise, but which I think belongs within the study of the formal sciences, not empiricism per se. Mach and James avoided the linguistic turn by pre-dating it, and Russell, it seems, had wearied of linguistic analysis by the time of his neutral monist period, when he was headed in a naturalistic direction.

4. Realistic empiricism proposes a working "umbrella theory" of the sciences, or a theory schema for constructing empirical theories in physics and perceptual psychology. This theory schema is an engine of

analysis for eliminating extraneous content, or mental imagery, from science and predicts the general form of empirical theories, just as specific theories predict data. The theory schema is a naturalistic philosophy continuous with science, but at the general, metascientific, level of theory design, and does not pronounce on specific empirical questions, or claim to be a first philosophy. Any value for the schema has to come through the specific theories it predicts.

There are three crucial historical stages in which realistic empiricism advances beyond traditional empiricism: (1) Mach's broadening of sensation to the notion of the neutral element which is also something physical (Chapters 1 and 2); (2) James's push beyond restricted empiricist epistemology to a directly realistic perception of mind-external objects (Chapter 3); and (3) Russell's adoption of neutral event particulars as a common basis for constructing physical space-time and sensory manifolds in psychology (Chapter 4).

Neutral elementary events

Realistic empiricists hold that the natural world is made up of individualized events embedded in real causal–functional relations to each other. These events and causal–functional relations are what really exist, and the rest (objects, extended bodies, fields, space-time, brains, and minds) are constructed out of them. These events are called "elements" by Mach, "event particulars" by Russell, and "pure experiences" by James. I will use the terms "elementary event" or "element" to cover them all.

Elementary events constantly change and are immediately replaced by others, so we must distinguish them from enduring objects, element-types, or properties. For example, one can call "John's stroll" different every time he takes it, or one can describe "John's stroll" as a repeating type of event that takes place every morning at 9 a.m. Elementary events are absolutely unique and non-repeating. Mach famously said that "nature has but an individual existence" (1883/1960, p. 580). For the realistic empiricist, as for the traditional empiricist, existence is always of *particular* matters of fact, and whatever exists at all does so as a concrete particular *event*, nature being just the sum total of these events and particular causal relations between them.

I think that events in realistic empiricism are further to be understood as the manifestations of dynamical powers and I think all three original authors do actually make some commitment to this view that events are caused by powers. This is most clearly stated by Mach, whose elements were akin to the manifestations of "potential differences" or forces in physics (see Mach 1883/1960, pp. 598–599 and Banks 2003, chs. 3, 7, 9). But Russell, too, speaks about his event particulars as the disembodied "interactions" or "effects" of objects and observers on each other (Russell 1921, pp. 101–102), and James talks about the "energetic" relations between his elements of pure experience, including mental events and "mental work" which he sees as no different from physical work or energy (James 1977, pp. 181, 289). Mach, James, and Russell thus made it very clear in their writings that elementary events are dynamical and forceful, or that these events *are* the manifested effects of causal powers of some sort. In realistic empiricism, nothing simply happens; every concrete particular event in nature happens because something makes it happen—or prevents it from happening. Events are always the concrete "token" manifestations of powers, including powers that completely block or equilibrate each other, seemingly leading to no effect at all, as in seemingly stable objects or force-free trajectories.

Since this theory is not about the linguistic meaning of the term "event," we need not accept that everything a philosopher might call an event really is one; nor do naturalists need to canvass all of the possibilities for what the word "event" might mean to a philosopher of language or an analytic metaphysician reconstructing linguistic usage. The natural-language description of events and powers and manifestations is usually superficial and must be broken down until the real natural powers manifesting in the event are identified, in the potentials and forces of the physical world, as Mach suggested. Indeed, I believe we can eventually drop the talk of "powers" and "manifestations" and speak directly about individualized physical potentials and manifestations of force in Chapter 6.

Finally, in realistic empiricism, powers are directly manifested in events by their concrete *qualities*, or what Mach actually calls 'physical' qualities (1905/1976, p. 15). Every natural event exhibits these qualities, or what Russell calls the "intrinsic character" of matter. These qualities are not mental, nor are they an extension of mentality to the rest of the universe, as in panpsychism. Qualities are simply the concrete empirical manifestations of powers in events, observed or not, that occur around us all the time. They are not restricted exclusively to the qualities of our sensations, which are very special and complexly configured natural events in the human brain. This idea, that mind-independent qualities occur in physical events, which seems so strange at first, can actually be found throughout much of the history of philosophy, even in such hard-nosed works as Moritz

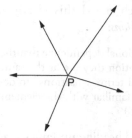

Fig 0.1   A physical point-event and its individual qualities

Schlick's *Allgemeine Erkenntnislehre* (1925/1985, p. 284),[1] Herbert Feigl's essay *The "Mental" and the "Physical"* (1958), and in contemporary philosophy of mind (Maxwell 1978, Lockwood 1989, Unger 1999, Chalmers 1996, 2002, Stoljar 2001, Rosenberg 2004, Banks 2003, 2010, and many others).

As Russell held, these manifested qualities serve to identify individual events. Elementary events are identified one by one, as particulars, and do not require further anchors, whether objects, properties, or universals, for their identity conditions (Russell 1947/1997). In Chapter 6, I will reconstruct these events and qualities using Grassmann's point-algebra, where the events are treated as points and the various qualities are like "spokes" sticking out of the point in a variety of directions, representing interactions from each possible causal point of view one can take on the event (see Figure 0.1).

Realistic empiricism is thus a kind of physicalism, with a description of natural events in terms of powers, manifestations, and concrete qualities, completely continuous with natural science. Within this enhanced view of the physical world, realistic empiricists offer a solution to the problem of how the realm of human experience (colors, pains, sounds) relates to physics. Russell called his view "neutral monism" meaning that neutral elementary events should be taken *neither* exclusively as mental *nor* as physical, at least not in the customary sense of the "physical," which

---

[1] In the quoted passage Schlick admits an array of natural qualities associated with individual events, but he says they are of no interest to science, which concentrates on the formal structure of events and not on individual events themselves, of which he thinks we can have no knowledge. He *then* adopts the view that phenomenal qualities are only met with in human observation and are therefore epistemologically privileged, in apparent contradiction to what he has said before. I do not seek to unravel this knot, but I think it shows that Mach's neutral view of sensations and physical events has already been lost. Schlick does not even see it as a possibility.

excludes psychology. Mach stated this clearly in a well-known passage from his *Analysis of Sensations*:

> It is only in their functional dependence that the elements are sensations. In another functional relation they are at the same time physical objects. We only use the additional term "sensations" to describe the elements because most people are more familiar with the elements in question as sensations. (Mach 1886/1959, p. 16)

Similar quotes can be found heading up James's radical empiricist essays and Russell's *Analysis of Mind* (see Banks 2003, chs. 7, 9). According to Mach, who made the initial breakthrough, the same event in the brain (**s/e**) can be considered a mental event (**s**), by associating it functionally with memories, mental images, trains of association, and other psychological variations, but it can also be considered a physical event (**e**) by relating it to physical objects by physical laws, such as those of brain physiology (see Figure 0.2).

Just as there is no inner dualism in the elements themselves, there is no dualism of functions or variations either. Even the terms "psychological" and "physical" only refer to provisional differences in the variations which happen to fall under different departments of study. For example, the sensation of a freely falling red ball may participate in the following variations

(M) All strong red sensations x are followed by green after-images y

which may be of more interest to those studying the human nervous system, than another variation

(P) All physical objects x freely fall in the visual field at 9.8 m/s^2

which is explained by the physics of free fall, and not at all by the organization of the sense organs and the brain. A sensation (**s/e**), however, can be substituted for x in both orders indifferently and it obeys both kinds of variations. What we explore in science and philosophy, therefore,

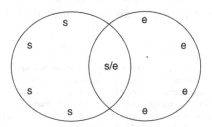

Fig 0.2   Sensations and elements

according to Mach, is events that belong to the whole unified fabric of experience-reality and not just one department of it. We can thus call the elements and variations "natural" or we can even call them all "physical," in anticipation of some future science of what the physical world really encompasses: sensations *and* physical events in one realm. So, unless I am specifically separating psychology from physics, I will use the term "physical" in this broader sense of a physics which includes psychology. Russell also does this quite often, as we shall see below.

The conscious ego

Mach and James both emphasized the fact that the red patch of a sensation does not get the quality of being red from an act of awareness or representation by the conscious mind. The red patch is also red in its purely physical variations, which do not involve the psychological variations of memory, attention, perception, and so forth. The red patch is a physical event in the human nervous system, and would remain red even if the patch were not being attended to consciously. For Mach and James, the conscious ego is regarded as a second-order "functional" connection among sensations, mental images, feelings, and other phenomena, and has no independent existence as a substance or stage or embedding circumambient medium, required for these other items to exist. For Mach and James, what we seem to be aware of in the unity of consciousness and its many acts is simply the unity of the many different functions carried out in our nervous systems, the composite result of levels of unconscious processing (see especially Ratliff 1965 and Banks 2001), and not some kind of mental theatre in which the contents achieve consciousness simply by being viewed by the internal spectator. Instead, what happens is that already-present contents become conscious by being functionally related to other contents such as mental images in functions of memory, time and space perception, judgment, and imagination. Or as Mach puts it in *Knowledge and Error*:

> Consciousness is not a special mental quality or class of qualities different from physical ones; nor is it a special quality that would have to be added to physical ones to make the unconscious conscious ... A single sensation is neither conscious nor unconscious: it becomes conscious by being ranged among the experiences of the present. (Mach 1905/1976, pp. 31–32)

Hence the ego delimiting mental from physical events simply does not exist, beyond a collection of functions ultimately realized in the physiological activity of the brain. As Mach provocatively put it, "the ego cannot be saved" (*Das Ich ist unrettbar*).

I suppose this is a major difference between realistic empiricism and many contemporary views of consciousness, including phenomeno-logical views that derive from Franz Brentano (see Smith 1994, Harman 1990). James and Mach were both very harsh critics of Brentano's notion of the "intentional inexistence" of objects embedded in consciousness. Russell was only able to become a neutral monist in the style of Mach and James when he gave up on the idea that there had to be a funda-mental, irreducible relation of acquaintance between a mind on one hand and its contents on the other (Russell 1959/1997, pp. 134–135). Once Russell realized such an act was neither necessary, nor introspect-ively observable, he abandoned it, although as we will see in Chapter 4, the issue is complicated by Russell's theory of knowledge, which remained a kind of representative theory of mental images unlike the naturalistic causal theory of "knowledge and error" held by Mach and James.

Functions and causal–functional dependence

Mach and James both strongly emphasized the fact that the causal–functional relations between the elements were as real as the elements themselves. The most basic sort of link is the causal relation between a particular token event and another particular token event. If we think of events as empirical manifestations of powers, as I suggest we should, it is natural that these power manifestations will affect and constrain each other and that this interplay is what is manifested in their causal relations to each other. So what we end up with is an interconnected causal mesh of events related to one another in real complexes, not isolated atomic occurrences associated with one another loosely or sub-jectively, as in the traditional British empiricism of Locke, Berkeley, and Hume.

Mach, James, and Russell all held this kind of dynamical view of their elements and, consequently, they also asserted some kind of explicit principle by which the elements are bound to each other in causal relations, or "causal laws" as Russell says. Mach asserted a general principle of the "functional, reciprocal dependence of elements on each other" (1872/1910, pp. 69–71; 1883/1960, p. 604) and James even defined his radical empiricism as entailing a belief in real, particular relations between events:

> The relations between things, conjunctive as well as disjunctive, are just as much matters of direct particular experience, neither more so nor less so than the things themselves. The generalized conclusion is that therefore the parts of experience hold together from next to next by relations that are themselves parts of experience. (1977, p. 136)

Hume claimed never to directly observe any connection between his impressions. In realistic empiricism, we observe the connection and the causal dependence between events directly, even if we cannot predict which particular event will occur next to, or after, another particular event. Mach declared that his elements *always* occurred bound up in functional complexes, and that elements were mutually dependent on each other. They were not little atomic sensations or sense data. Mach adopted the neutral language of "causal–functional" dependence to replace the language of causation, which he thought was a relic of the mechanical philosophy of nature, and implicitly contained the assumption that a mechanism of some kind was responsible for every natural regularity or dependence. Mach's elements (Mach 1896/1986, pp. 328–329), as he says, simply 'abut upon each other' without any underlying support or mechanical system somehow lurking behind the manifested events. Thus Mach did hold to a robust realism about relations and real causal dependencies between elements; he just did not *call* it causation.

More must still be said about this peculiar language of "causal–functional" connections or variations intended to replace the mechanistic causation that Mach so deplored. Mere "mathematical functions" are notoriously vague since anything is *some* function of anything: the price of bread and rising water levels in Venice. There has to be a way to restrict the possible functions under discussion while leaving enough flexibility for different kinds of function to exist over the same elements, the two psychological and physical orders for example, but also micro- and macro-levels of functional dependence, polyadic dependences, and even a variety of different ways to order the same physical events into objects, law-like regularities, or perhaps even abstract tabular or matrix-like arrangements, as we shall see below. But too much latitude and anything goes. This is why it is so essential to see the elements as manifestations of powers, and their individual causal connections as the real effects one event has on another. Particular causal–functional connections are grounded in the causal behavior of the events themselves and their qualities, and are not subject to arbitrary choice (see Banks 2004).

It is, of course, very rare that we can isolate an "elementary event" from the rest of experience. As Mach says, nature "does not know" elements of experience but only complexes, and the vast majority of the time we, too, are dealing with complexes and objects which are epistemologically basic for us, not the individual elements and functions which make them up. The actual evanescent elements and their qualities that occur around us are too fleeting to be apprehended directly, and they are quickly replaced by new ones. It is, rather, the repeating patterns or functions that make up the objects, laws, and systems present to us in experience, not atomic "given" elements. For this reason, the functional relations, at least in the physical world, must be very strong and permanent, equal in durability to any object. Someone might scoff at that by pointing out that the Himalayan mountains are *objects*, not flimsy functions of still flimsier evanescent events. The human mind seems to grasp out instinctively for permanent objects, an instinct Mach especially tried to discourage. But, as Mach and Russell both insisted, what *is* an object that would make it any stronger than a permanent law or function? For them, the solid, causal–functional law, or invariant function, of the elements does not require the further existence of a substance, or substratum, in which to anchor effects or qualities. Elements are already anchored as well as anything could be simply by abutting upon each other causally and they are grounded directly in their manifested individual qualities. These functions cannot be just formal or arbitrary combinatorial relations. Rather, it seems the function expresses some kind of embedded relation realized in the causal behavior of the elements.

We have, just as Mach and James both insist, *real* functions embedded in the manifested qualities of the elements, and their causal behavior, and backing up the solidity of objects, which is not present at all in Humean empiricism with its passive sensations and mental associations of ideas, or in the concocted "remembered similarity relations" of Carnap's *Aufbau*, which are obviously subjective and mentalistic. As I have pointed out before (Banks 2003, pp. 10–12, 2013a) there is thus no true philosophical similarity at all between the elements and functions of Mach or Russell and positivistic constructivist projects like the *Aufbau*, which are much more traditionally empiricist in spirit. It is true that Carnap, Schlick, and Neurath considered a physicalistic protocol language (see Uebel 2007 on this point) and it is well known that Carnap even considered a parallel physicalist version of the *Aufbau*, but the Vienna Circle thinkers remained dualists about the conflict between a "physical" language and a "phenomenal" language for observation reports, or *Protokollsätze*, and never adopted a truly neutral view à la Mach, for whom the problem simply vanishes. Hence the whole *Protokollsatz* debate took place on a much lower level philosophically.

Mind-independent world elements

A question that arises naturally is what to do about completely mind-independent events and their qualities. If any sensation (s/e) can be interpreted in another context as a physical event, does it follow that all elements are also sensations (e/s) under some interpretation? Some will still defend that reading, but it seems to me to be far too phenomenalistic, as I will show. Should we then admit other elements (pure "e-elements") that are not anyone's sensations, and not even anyone's *possible* sensations? This question has plagued the literature on Mach, James, and Russell and was a major cause of confusion at the time when they first advanced their positions, but I think it is now possible to give a definitive answer. For Mach and Russell, the sensation-elements (s/e) are only a *special class* of the mind-independent e-elements, or what I have elsewhere called "world elements" (Banks 2003). There is no question, for either author, that the vast majority of events in nature do not occur in a human nervous system and there is no way to represent the external world, or our theories about it, as a mere "catalogue" of present human sensations or even purely observable contents of scientific theories. Russell was quite explicit about the existence of mind-independent "sensibilia" at one point in his career, when he still believed in the theory of acquaintance, and later just neutral "event particulars," after his neutral monist conversion. Mach and James tended to emphasize that (s/e) elements were *already* physical events causally linked to other physical events, and that the barrier had already been breached in the realm of sensations, which are already something physical, but they both advanced further still.

Mach often spoke of the need to "add elements in thought" (*hinzudenken*) in order to connect up the fragmentary experience of our sensations into a real experience of objects in space and time. Even when we observe the path of a parabola, for example, we add in thought the past and future stages to what we presently sense in order to complete the experience. We likewise add unobserved backs and sides to chairs and other objects to fill them in as three-dimensional objects in perspectives, even if no observer is presently situated at those other perspectives to observe them, or could even possibly occupy them all at once from an egocentric perspective. And of course we really mean when we look at a chair, and not a collection of colored blobs, that those missing perspectives are occupied by real, but unobserved, events, and not our mere "mental" additions in imagination and memory, which would have no force to affect events in nature and would not physically provide the sides and back for a solid object in front of us. In his *Science of Mechanics*, Mach gave the example of a vibrating rod in a vice (1883/1960, p. 587). We observe the vibrations of the bar at one end and follow them down to the other end of the bar, where they become invisible. Does anyone seriously think the bar simply stops vibrating just when we can no longer see it? Or that merely mental or imaginary additions of elements could keep the invisible end of the bar vibrating? It is absurd to think we cannot connect our experiences continuously with other elements that are not sensed by us, especially when we have a rule or a function to guide us.

> When we mentally add to those actions of a human being which we can perceive, sensations and ideas like our own which we cannot perceive, the idea of the object that we form is economical. The idea makes experience intelligible to us; it supplements and supplants experience. Now, this is exactly what we do when we imagine a moving body which has just disappeared behind a pillar, or a comet at the moment invisible, as continuing its motion and retaining its previously observed properties. We do this so that we may not be surprised by its reappearance. We fill out the gaps in experience by the ideas which experience suggests. (1883/1960, p. 587)

Of course, Mach did sometimes use the word "sensation" when he seemed to mean "element," and there are passages where he speaks about nature (Mach 1883/1960, p. 579) and the visual world as if they "consist only of our sensations" (Mach 1886/1959, p. 12), but it is very wrong to think Mach did not also believe in mind-independent elements needed to complete our experience of objects. As Mach knew, we are compelled to "complete" our sensation-elements outward, adding new particulars, just as we complete partially filled-in functions *within* experience. Indeed there is some evidence that Mach actually saw the physical universe as one connected fabric of experience, which included unobserved elements within the notion of experience. The crucial issue then is whether external elements are causally *continuous* with observed experience, not whether they are directly observed.

The real absurdity, Mach thought, was to postulate objects like Kant's *Ding an sich* that could not be causally connected with experience by any means. For most of his career, Mach actually thought atoms were *Dinge an sich* (see 1883/1960, pp. 588–589) for interesting physical reasons having to do with the problem of spectral lines, but he recanted when the continuity of atoms with experience was established (see Blackmore 1992, p. 151, Banks 2003, pp. 12–13). Mach's objection to the *Ding an sich* is quite cogent, however. Consider an object that reacts with a series of different environments and produces a series of manifested effects. Suppose that some of the behavior of the object is stable over all of these different interactions. Is it right to say that this object would still have these manifested properties intrinsically, even when it is not interacting with anything else? Mach says no, these sorts of isolated substances (1883/1960, p. 589) with "intrinsic" non-interactive properties should not be admitted in science. No interactions, no objects.

We are justified in the more limited view that behind experience we simply find *more* experience, figuratively speaking, more events and inter-actions in continuous functional relations to the sensation/elements of the sort that experience already presents to us. For Mach, science is the study of this whole connected fabric of experience-reality, not just one part of it. This is what he meant by emphasizing that experience always needs to be *completed* in thought by adding more than just the observed elements. It is totally inadequate to present Mach's views as an economical catalogue of present sensations, completely ignoring the role of real, grounded func-tions and elements "added in thought," as still occurs much too frequently in the literature (for a closer reading of Mach's economy of thought see Banks 2004).

In keeping with his scientific world view, Mach said that the elements were those individual events in nature which "we cannot divide further at present," "provisional only like the elements of alchemy," (Mach 1905/1976, p. 12n) and he refused to pronounce definitively on their nature. Were they like little energy quanta, tiny spring-like entities? How do we describe them in legitimate physical terms? Feyerabend (1984, pp. 19–20) wrote that for Mach, "elements were more fundamental than atoms." As we shall see in Chapter 1, despite some strong suggestions that the elements are expressions of potential differences, Mach, the physicist, never specified the physical nature of his elements, in keeping with his view that inquiry must always be kept open. Russell, however, defined his event particulars as "what common sense regards as effects or interactions of objects," in his example of breaking up the star Sirius into the sum of its effects on its environment:

> According to the view that I am suggesting, a physical object or piece of matter is the collection of all those correlated particulars which would be regarded by common sense as its effects or appearances in different places ... if photographs of the stars were taken at all points in space

throughout space, and in all such photographs, a certain star, say Sirius, were picked out whenever it appeared, all of the different appearances of Sirius, taken together, would represent Sirius. (1921, pp. 101–102)

Certainly, then, these robustly physicalistic elements do seem to manifest natural powers and they become the subject of physical–philosophical investigation. They cannot be "given" in any foundationalist epistemic sense without making the whole idea of a further investigation of them ridiculous. That sort of approach came later with the Vienna Circle philosophers, whom Feyerabend (1970, pp. 179–181 and 1984) accuses of misunderstanding and perverting Mach's original elements for their own purposes, for example Schlick (1925/1985), who sees Mach's elements and functions entirely in phenomenalistic, verificationist terms (see also Uebel 2007). Another low is reached in Rudolf Carnap's essay "Empiricism, Semantics and Ontology" (1950), where he says that the question of what exists is to be reinterpreted as the pragmatic question of which "thing-language" to adopt. Mach would never have said this, for his elements are real events, not linguistic constructs or conventions, or nodes in a logico-linguistic structure as they seemed to be for Carnap. As Michael Friedman (1999) points out, Carnap was not actually interested in the elements of experience, or nature. He wanted to isolate the formal, structural features of science from its empirical content, an entirely different project. But, as I repeatedly emphasize (especially in Banks 2013a) the purpose of science and philosophy for Mach, James, and Russell was to investigate reality, not how we *talk* about reality, or second-order claims about the language or structure of science and its canons of methodology, models of explanation and so forth which have become familiar fixtures of logical positivist philosophy of science. As Mach succinctly puts it in *Knowledge and Error* (1905/1976, p. 84): there *are* no canons of method-ology, such as Mill's methods, which he says "would not lead beyond the most rudimentary results" (1886/1959, p. 92), nor are there schematic models of explanation, beyond scientific theories and models themselves, which are often replaced by new theories of precisely the same form. In short, the *same* methods that lead to knowledge under some circum-stances also lead to error under other circumstances and only experience can be the judge:

> Knowledge and error flow from the same mental sources, only success can tell the one from the other. A clearly recognized error, by way of corrective, can benefit knowledge just as a positive piece of knowledge can. (Mach 1905/1976, p. 84)

The twentieth-century movement toward second-order logico-linguistic and or structural questions is exactly the move Feyerabend (1970) condemned by calling philosophy of science "a subject with a great past"—in the "great old days" of Mach, Duhem, and Poincaré, before the rise of Wittgenstein, Vienna Circle positivism, and linguistic analysis. According to Feyerabend, the whole retreat away from the real and into a second-order analysis of language and methodology, and highly artificial models of scientific explanation, is completely alien to this earlier movement, which dealt with first-order philosophico-scientific questions as part of one unified area of study.

So, to sum up this section, the realistic empiricist elementary events are clearly not "atomic sense data," à la Hume, nor are they the subject matter of the *Protokollsätze* of the Vienna Circle. They are real, concrete, dynamical events in the world, the same as those described by natural science. Moreover, elements à la Mach are only isolated provisionally from their functional complexes by the method of variations and are not directly given. Epistemologically, it is the objects or complexes that are the primary things in the actual investigation, not the individualized elements. And for other reasons, to be given later when we discuss James, elements themselves would not provide representative knowledge of anything outside themselves; they simply exist. So there is no remaining sense in which the elements can serve as the primary givens for a traditional empiricist view, or any kind of sense data foundationalism.

The umbrella theory of elements and functions

Methodologically, what we have inherited from Mach, James, and Russell is, I suggest, an overall "umbrella" framework to work with and fill in, consisting of dynamical elements and causal-functional variations grounded in the causal behavior of those elements. An umbrella theory is a schema or pattern for designing or even predicting specific theories which must be discovered and verified empirically. A paradigm example of an umbrella theory is Darwin's 1859 theory of evolution, which is really a schema involving the factors of variation, selection, and reproduction. Any particular evolutionary story will indeed have these overall features, in some form or other, and in that sense the 1859 theory is indeed predictive of future detailed *theories*, but the schema on its own is *not* predictive of any specific story or line of descent (as Fleeming Jenkin 1867 famously pointed out to Darwin). Yet as the theory of evolution developed into its modern form, a *specific* evolutionary story was advanced for each life form and then confirmed with converging but independent lines of evidence from the fossil record and geology, from DNA gathered from living descendents, as well as evidence from embryology and epigenetics. What umbrella theories really do, then, is predict *theories*, the same way that theories unify and predict empirical laws and data.

The realistic empiricist umbrella theory gives a schema for developing particular empirical theories based on element-and-functions. But, as in all umbrella frameworks, if the elements are left completely general then *anything* is an element and the theory schema comes out empty. If *any* function can be introduced to associate the elements with each other, then *anything* counts as a causal–functional variation. The umbrella theory would thus predict any and all theories, unless more is added. We already said that elements are individual events, not recurrent types of events or universals, like properties. They are not objects, but rather comprise objects by being embedded in particular functional relations of various sorts, particular-to-particular links, which are as much a part of experience-reality as elements themselves are. We also said that elements are dynamic and powerful: they fall naturally into a network of causal-functional relations by exerting reciprocal effects on each other. The causal–functional relations are grounded in this network of effects and in the individual qualities of the events in which their causal powers are directly manifested. This much, I believe, can also be suggested by the works of the three original authors.

The metaphysical argument for this kind of element-centered view would be that everything that concretely exists in nature is a particular *event* and every event is the manifestation of some kind of causal power. Things don't merely exist; they happen. And they don't happen unless something makes them happen—or prevents them from happening. Force and action, not matter, is the true essence of our world. This idea, that reality is the manifestation of power, can already be found in the Pre-Socratics, and is attributed to Plato's Eleatic Stranger in the *Sophist*:

> My notion would be, that anything which possesses any sort of power to affect another, or to be affected by another, if only for a single moment, however trifling the cause and however slight the effect, has real existence; and I hold that the definition of being is simply power. (Jowett 1871)

Today, we can speak more precisely by using the language of powers and manifestations, about which there is a considerable literature to draw upon (for example: Mumford 1998, Martin 1993, Heil 2003, 2005). We describe an event as the token manifestation of a natural power under specific circumstances or conditions of manifestation. If other factors get in the way, blocking, changing, or masking the effect, then these are *also* powers and can be included in a more complete description of the event, or the frustration of an event.

Later, I will further enhance this view by assuming what is called in the literature the *token identity thesis* between powers and the token manifestation of those powers. An exercised power is "token identical" with its particular manifestation, or as Hume once put it, there is no distinction between a power and the exercise of it. According to this thesis, a power will be identical with its individual manifestation case by case, and there is no power in itself underneath, nor does there need to be, in order to assert the identity of the same power exercised over a variety of different token manifestations. Powers, or potentials, contribute the objectively real structure of the world persisting across particular token events in places and times, but token manifestation events account for the particular existence of the world at specific places and times.

What makes my view an empiricist view is the denial that there must be any a priori conceptual relations between the various token manifestations of powers in different circumstances. Although a power is identical with its different manifestations this identity between the power and its different token manifestations can only be known a posteriori, empirically, and cannot be deduced from concepts alone. The identity of powers and their token manifestations in events is then, as philosophers say, an a posteriori necessary identity. These issues will be explored more fully in Chapter 5.

Finally, in Chapter 6 I will suggest we do away completely with the philosophical language of powers and their manifestations and move to the actual physical language of individualized potentials and forces governing natural events. These additions in the last two chapters are of course my own attempt to reconstruct realistic empiricism in a contemporary vocabulary. I realize some will object to the reconstruction as not keeping faith with the original, while others will have problems with the concepts and arguments of the reconstruction on its own terms as a piece of philosophy, and I would not want Mach, James, and Russell to be tarred with the same brush along with me. However, I do feel strongly that the view I represent *is* a natural continuation of the Mach, James, Russell view, like a direct descendent of an older ancestor.

Elements and qualities

One of the most difficult aspects of realistic empiricism, of course, is the idea that all elementary events, whether they occur in the brain or not,

exhibit qualities as the concrete manifestation of their powers. This idea
was the greatest barrier to my understanding of the view because I thought
that admitting the omnipresence of natural qualities would lead to some
form of panpsychism, in which the physical becomes imbued with mental
properties all the way down to atoms and particles. I now see that the
enhanced physicalist view of the world involving events and natural
qualities comes first, *before* the mental in every sense. *All* individual natural
events involve manifestations of concrete qualities, and sensation qualities
are just the higher-order qualities of very complexly configured events in
our brains. I will claim that sensations are events that collectively manifest
the *configured* powers of individualized cells all acting together in a combined way. But this is not so very different from what goes on in other
natural events. The idea is not to think that physical objects are actual
bundles of human sensation, nor must physical events and qualities be
similar to human sensations, since we are not creating a line of descent
from sensation qualities to the qualities manifested in physical events, as
panpsychism does.

Clearly when Mach, James, and Russell say that sensations (**s/e**) are
already fully-fledged physical events, qualities and all, we are being asked to
think of the sensory quality of that event, like blue or green, as a physically
real occurrence with a physical explanation, in an enhanced sense of the
word "physical" that includes psychology. Russell (1927/1954) proposed
that *all* natural events, observed and unobserved, manifest qualities when
they occur, in addition to their more abstract quantitative properties and
relations. Traditional physics measures the external relations of bodies and
motion, but according to Russell natural event particulars also possess an
"intrinsic character":

> A piece of matter is a logical structure composed of events; the causal laws
> of the events concerned, and the abstract logical properties of their spatio-
> temporal relations are more or less known, but their intrinsic character is
> not known. Percepts fit into the same causal scheme as physical events and
> are not known to have any intrinsic character which physical events cannot
> have, since we do not know of any intrinsic character which could be
> incompatible with the logical properties that physics assigns to physical
> events. There is therefore no ground for the view that percepts cannot be
> physical events, or for supposing that they are never copresent with other
> physical events. (1927/1954, p. 384)

Russell's idea is not to psychologize the physical by making everything
some kind of mental event: it is exactly the opposite, namely to subsume
psychology within physics by *enhancing* the content of the physical world,

giving it a rich, qualitative grounding in the intrinsic character of its natural events, while leaving its structurally described exterior intact. We get an enhanced view of the physical and a grounding of its abstract quantitative structure in natural qualities, and we get an explanation of sensations and experience in the bargain, so Ockham's Razor is satisfied and the assumption of individual qualities for events does explanatory work that neither the physical nor the mental alone can match, for physics just leaves out sensory experience, and its own grounding for its relational structure as well, and psychology leaves out the grounding of mental phenomena in the physical brain. If Russell is correct, then the philosophical meaning of "quality" must be changed from its seventeenth-century meaning, restricted to the "secondary qualities" of our sensations, to a more robust physicalistic meaning, suggesting concretely manifested effects in any and all natural events, which is perhaps what "quality" would have meant to Aristotle or Duns Scotus: some kind of qualitative manifestation of a power which has nothing in particular to do with being perceived by minds.

Elements in physics

Surely if the Machian–Russellian elements are to play a role in the unification of science they must also have a clear underpinning in physics, the most fundamental natural science. The elementary events and their individual qualities must do some explanatory work, else they will be a mere unnecessary add-on to physical theory. We should thus begin by developing what I call an "enhanced physicalist" view in developing a theory of elementary events within physics.

Following Mach closely here, physical events are the manifestation of physical forces, and forces are manifestations of a potential difference of some sort. These are the fundamental physical happenings: a particle, like an electron, moves across a potential difference, like the energy levels in an atom, and work is done, a photon is emitted and carries away the energy developed. A loop of wire cuts across the equipotential tubes of a magnet and an electromotive force exerted by an electric field pushes the charges around in the wire. It is true that we only know about potentials because of the forces they manifest. An absolute potential level is not observable, only potential differences manifested by forces. In that sense, the physical idea of a potential does neatly encapsulate the vague philosophical notion of a "power" and force is the natural expression of a "manifestation" of power under a given circumstance.

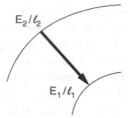

Fig 0.3   A potential jump

Mach expresses this quite well in his writings (see Mach 1886/1959, pp. 351–352, speaking of potential "differences" as driving all natural change and events; Banks 2003, chs. 10, 14) and comes very close indeed to simply calling his elements the manifestations of potential changes as forces, say a certain amount of energy (E) transferred between equipotential levels l_1 and l_2: $E_2/l_2 - E_1/l_1$ (see Figure 0.3).

Mach believed that the language of potential changes served to explain natural events directly and was far more general than was understood at the time. For example, he sought to free general natural laws like the conservation of energy or the principle of least action from any realization in terms of a mechanical system of particles and central forces in space and time, and then finally from any microphysical realization or models whatever (Mach 1883/1960, pp. 598–599; 1896/1986, pp. 328–329). The idea here is not to confine ourselves only to observable events and ignore the actual natural world and processes underneath, but to isolate what Mach sees as ultimately underneath all mechanisms and models, even in mind-independent nature. Even if the curtain of observation could be drawn back, behind it we would still find natural events and their relations to each other, even underneath the unobserved objects or mechanisms if such there be. This is how Mach's remarks against mechanical physics and in favor of "phenomenological" physics should always be understood, never as an injunction to confine ourselves *only* to the phenomena of human observation and nothing else.

Some suggested developments: algebraic techniques

In my view, we can develop Mach's ideas further in this direction by defining a physics tailored to individualized events (Banks 2008, 2013c). In physics, we think of there being only a small number of truly irreducible potentials, corresponding to the fundamental forces of nature. But the language of potentials and forces is all-purpose, and there is no reason why we cannot go in the opposite direction and define *individualized* potentials of all sorts and combinations, for situations that might arise empirically, a different potential for each individual event and its manifested quality from each different causal point of view on the event. We can have as many potential transitions as we have events and causal perspectives on those events. We don't usually give such a detailed analysis because we can assume that an event is the same no matter who is observing it from where. But for truly individual events, we can estimate the potential differences they represent from a variety of different points of view, any of which can serve as our zero, or center of causal perspectives. For example, if A sees a particle jump from O to A, then B will see the "same" jump indirectly as occurring from B to O, O to A, and then A to B. The actual jump made in the first event can be represented from a different vantage point as a different jump which, if it *had* happened B's way, would have had the same overall result. A variety of different causal perspectives on an event can be obtained by changing the position, angle, distance, velocity, or time of observation by the second observer, all of which we consider to be different causal perspectives on the event which manifests different individual qualities to him and, in a sense, is a different event, rotated into another point of view.

To develop these ideas further, I will introduce a model based on Grassmann's algebraic point-calculus (Grassmann 1844/1955, Banks 2008, 2013c, Browne 2009), which provides a natural, and coordinate-free, way of understanding individualized events and their qualities algebraically in a space of perspectives, prior to metric determinations or even spatial concepts. These point-events live in a kind of abstract "quality space" useful for classifying them by their individual qualities, or what is the same thing, their jumps from all possible zeroes. In quality space, an event is classified by the qualities it manifests, not by its position in space and time, or by relating it to an underlying extended physical system. Individualized potential jumps from a given zero point to a point-event can be represented by vectors in the quality space (see Figure 0.4).

An individual event is represented as a point, and its qualities are represented as the variety of jumps to the point-event from different zeroes, as it would appear to different observers with different origins. Each zero represents a different observer's causal perspective on the event and the jump is represented as it would appear from his point of view only. Two events alike for one zero can always be further differentiated by adding all other jumps from all of the other zeroes. The variety of potential

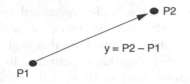

Fig 0.4 A point-event as a jump from a zero

jumps serves to identify the individual event, each 'spoke' representing a manifested quality of a different type from a different point of view on it. The individual qualities of a particular event will be different from different perspectives. Change your zero and you change not only your perspective on the event, but you also change the quality the individual event manifests, or would manifest, from that perspective. Assembling all of these different points of view on an event gives us an objective multi-perspectival representation of it from all points of view, expressing not only how it did happen but a variety of equivalent objective ways it might have happened. This identifies a point-event uniquely and objectively. Two point-events are considered identical if and only if they have all of the same individualized potential jumps from all of the same zeroes.

To a physicist, of course, this extra specificity may seem artificial and overly stringent, locking events into individualized perspectives, even changing their individualized qualities from one such perspective to another where physics would say these are just the *same* events and potential jumps observed from various angles, positions, velocities, causal perspectives, or in different observable bases. I hope to show that the added attention to detail allows us to ascribe a meaning to an event as a singular occurrence. This view, which takes the event as a concrete particular, as it actually occurs, and only from a certain limited perspective, is, I think, a natural development of empiricism, which has always emphasized the concrete particularity of existence. We are also following Russell, who constructed a system of individualized perspectives around event particulars, and who thought of this as the basic reality underlying the physics of particles in space and time (Russell 1913/1926, 1921, 1927/1954).

The pay-off of this event-centered view is that it allows us to build up extended physical systems from within an abstract quality space developed algebraically, using the mathematics Grassmann developed for this purpose, without assuming a background space, or space-time as Russell did. In the second stage, we want to move from the events in quality space, to the idea of an extended *quantity* capable of measurement from all different perspectives, such as a perspective-independent physical system. That is, we think of the physical system as an extension constructed directly from events and qualities and the characteristic "associative–dissociative" relations that trace out extensions, which we will introduce and explore in connection with Grassmann and his exterior product operation in Chapter 6. We then arrive at a measurable extension, which can be measured from a variety of different perspectives in direction, position, velocity, and time. Instead of using the philosophical language of "objects" and their intrinsic or relational "properties," more suited to natural language, we adopt a language of algebra (points, vectors) and invariants.

Following Grassmann's lead, I will show how it becomes possible to combine events algebraically into more complexly structured extended magnitudes and I will put forward this algebraic technique as a better construction of extension than Russell's because it does not assume the intuitive concept of extension as given in advance, but really does analyze it, revealing more structure behind it. That overall philosophical view, which begins with point-events in quality space and ends with extended objects in space and time, dovetails nicely with Grassmann's philosophical understanding of his own theory as an explicit conceptual construction of extension, as I will show in detail in Chapter 6.

Realistic empiricist epistemology

One of the pillars of empiricism has always been its account of human knowledge, going back to the title of Locke's *Essay Concerning Human Understanding*. Traditional British empiricism tended to focus on questions of human knowledge to the exclusion of the physical world. Yet those questions were in many ways the right ones: What do we actually know? What do we actually see happening? The problem was that the existence of the external world tended to fade into the background if there was no direct knowledge of it. Vienna Circle empiricism was well-nigh obsessed with the observational–theoretical distinction (see Uebel 2007). This unhealthy obsession forced the Vienna Circle philosophers to attempt rational reconstructions of scientific theories in which the theoretical T-terms had been reduced as much as possible to tortured logical constructions of the observational O-terms, distorting theories beyond recognition to fit a rather simplistic philosophical thesis (see Suppe 1977).

What is certainly right about empiricism is its emphasis on the actual concrete, individual events that occur in nature, its stringency, and its skepticism of speculative models and theories not firmly tied to observational consequences. Both Einstein and Heisenberg were able to spark scientific revolutions by re-asking those naïve questions: "Have we ever really observed synchronous, spatially separated events? Do we actually see the paths of electrons, or do we see instead the frequencies and intensities of the emitted light?" Clearly, these questions were only a kind of entering wedge for new theories, with many other components and structures that are not directly observable, for example the wave function or the gravitational field. But without a prior clearing of the ground, would there even have *been* a new theory? The point is often urged against empiricists that all facts are theory-laden, but a fresh, stripped-down view of the facts can also break the grip of older theories, naïve visualizations, or historically conditioned ideas which seem a priori and unassailable at the time. Moreover, when standards of stringency are too relaxed, and practically any statement can be held fixed, or revised, by adding more auxiliary hypotheses and assumptions, theories multiply out of control with nothing to constrain them. If we no longer even ask for verification, or treat it as naïve to do so, or refuse to accept that our theories could be falsified by experience, we are just asking for pseudoscience.

The fault of traditional and Vienna Circle empiricism was that they restricted their view of natural events only to those we can observe, and did not extend equal rights to all events, including those making up the mind-independent external world. A traditional Viennese empiricist would object: "but the elements of the external world are an added metaphysical assumption." To which we respond: "Yes, but you have already made metaphysical assumptions by *restricting* your scope to observed events and ignoring other possibilities." Even epistemology comes fraught with assumptions about what kinds of things we may safely assume, which always involves a prior metaphysics of experience. The traditional empiricist is right to take experience, even sensations, in a realistic sense, but he is too myopic in not realizing that our experience is just a fragment of reality, full of what William James calls a "concatenated" structure of real relations as well as unobserved events continuous with observation. Beyond experience are simply more of the same kinds of events and interactions that we encounter *within* experience, so we are not really making any bold leap beyond empiricism in considering events we do not directly observe.[2]

---

[2] There is a passage in, of all things, the 1929 Vienna Circle Manifesto, *Wissenschaftliche Weltauffassung*, in which the writer, probably Carnap, asserts that "science is all surface," perhaps in this sense of being "all experience" in the sense I mean it above, all events and interactions in causal–functional relations of the sort we experience. Unfortunately the passages which follow seem to suggest that Carnap meant this in a much more restricted sense, as if there were nothing but narrow human experience and observation.

But the traditional empiricist is right to be skeptical of extending our sensory conditions and naïve visualizations of the external world, as if it were merely a continuation of experienced objects. Why should our perceptions or imaginations be any guide to mind-independent nature, since naïve visualization can be deeply misleading? The history of science is littered with, now long-forgotten, subtle fluids and media, caloric, phlogiston, vortices, unobserved mechanisms like gear wheels and springs, luminiferous ether, and all sorts of models that made some true predictions and aided visualization, but turned out to be unnecessary or false; see Larry Laudan's (1981) gleeful refutation of naïve model-theoretic realism.

All of that is repudiated now, but this does not help us avoid similar mistakes in the future. In fact, a little more humility might be appropriate. Empiricism is correct to point out that not every regularity or law in nature requires models or gear-work mechanisms to explain it. Some regularities, like those of mechanics, thermodynamics, relativity, or quantum mechanics, may simply be taken as postulates of experience, needing no further realistic explanation. This divide between postulate- and mechanism-based science is sometimes illustrated by pointing out the amusing differences between so-called "English" and "French" science. Purportedly, Pierre Duhem and Henri Poincaré recoiled in Gallic horror when they encountered Faraday's and Maxwell's theory of electromagnetism, because it seemed to them to be an overtly visual model of the electromagnetic field, in "tubes of force" and unobservable "fluxes," instead of stark, abstract equations, in pure postulate form.[3] Of course, for every story in the history of science where a postulate is freed of crude visual mechanisms, there is another where the "crude visualization" advances the theory, for example when evidence converges from several independent quarters, eventually verifying the mechanism and confounding the positivist skeptics. Jean Perrin's 1907 convergence of evidence argument for the existence of atoms is a classic case. Ironically, twenty-five years later, exactly these sorts of convergence arguments *failed* to establish any inner reality for processes inside the atom and empiricism seemed to triumph once again over mechanistic realism as alternative, divergent but complementary, descriptions of natural phenomena replaced convergent ones.

---

[3] In defense of the English, the modern theory of differential forms gives an exact meaning to the speculations of Faraday about "tubes of force." A differential 1-form can be represented as a series of parallel planar surfaces representing equipotential levels and a differential 2-form can be represented as two intersecting stacks of planes into equipotential cells or tubes with an orientation, just as Faraday originally said.

How, then, are we to continue experience into the unobserved without falling into either trap, realism or positivism, and how do we preserve the highly desirable stringency of traditional empiricism without merely cataloguing observations and truncating science until it becomes unusable? In Chapter 3, I will suggest that James (and Kant) have already shown us how to do this, albeit with a little reconstructive work and some additions of my own (Banks 2013b). James advances a totally original argument for direct realism about perception, in which we do perceive the real proper parts of external, mind-independent objects in perspectives around the observer, and can assert such a judgment of perception even under conditions of skeptical uncertainty. His famous Memorial Hall example is the lynchpin of this argument, which I will reconstruct in detail in Chapter 3. However, we find that our perceptions of spatio-temporal objects are only one format, one possible representation that mind-external objects can exemplify. Other formats, other visualizations and models, are equally possible candidates for the objects of perception, and, as Kant suggests, would be perceived by other rational beings with different sensory systems and imaginations from ours. Thus chastened, we allow these mind-independent objects to be constructed in the same way as objects of experience, by elements and functions, assuming unobserved elements and functions to complete and construct the various extended spatio-temporal perspectives of objects. What we cannot do is assume that our construction is the *only one* describing reality. Certainly a continuation of experience into the objects of a model may be *one* permissible way to empirically represent reality, and we can even be, as Kant says, "empirically realistic" about it. But there may be other candidates equally suited and equally objective. Hence sticking to the methodology of elementary events and abstract functions is the proper minimal theory, which avoids the assumption of a specific and unique format for the representation of objects, in particular a spatio-temporal format. Realistic empiricist epistemology remains empiricist and direct-realist, but not naïvely realistic. This kind of Jamesian–Kantian realism is superior to traditional empiricism, but also to naïve model-making transcendental realism.

The historico-critical method

It is time to bring this long introduction to a close with some remarks on the historico-conceptual method used in the book. Most works in philosophy are either historical or contemporary, but the historico-critical method is twofold: it rediscovers and reconstructs concepts from the history of ideas, and it also engages in their contemporary analysis and redeployment. Some ideas in this book will be of purely historical interest, some will be reconstructed with contemporary vocabulary and additions of my own, and some will be both. The realistic empiricist tradition must be rediscovered in an historical way, since it is no longer well known or understood. But I will also advance my own version of the view. Both approaches complement each other and should not be separated artificially into two camps. As Mach recognized more than a century ago, a productive historico-critical discussion is often the best way to analyze contemporary problems.

CHAPTER I

Mach's physical elements

Introduction

The originator of the modern realistic empiricist position was Ernst Mach (1838–1916), the Viennese physicist and psychologist whose analysis of concepts in physics, in particular space, time, energy, and mass, became famous through his book the *Science of Mechanics* (1883/1960).[1] His results on the analysis of sense perception, along with his views on the relation of psychophysics to physics, were gathered together and published in his *Analysis of Sensations* (1886/1959). Mach's physical and philosophical views were highly influential in Central Europe and beyond, on philosophers and scientists, and directly inspired Einstein's early work on the theory of relativity (Einstein 1949, p. 7).

Mach often emphasized that he was not a professional philosopher and did not seek to found a school of thought. The task of philosophy, as he saw it, was to provide a unified view of the sciences: "I make no pretension to the title of a philosopher. I only seek to adopt in physics a point of view that need not be changed immediately on glancing over into the domain of another science, for ultimately, all must form one whole" (1886/1959, p. 30).

Mach's world view of "elements" contained both empiricist and realistic features. The empiricism is well known, and often criticized, but his realistic tendencies are not. Mach's physical elements have not often received the kind of attention they deserve in a well-rounded treatment, which I tried to give in my 2003 book, *Ernst Mach's World Elements*. This is probably because of a lingering association between Mach and the logical positivists of the Vienna Circle, whom he influenced, but who took his ideas in a distinctly different direction from what Mach originally

---

[1] The historical roots of the position go back at least as far as Kant and Leibniz, if not before, but tracing these roots would take us too far afield.

intended. Because of this historical association with the Vienna Circle, many still believe Mach was a naïve "phenomenalist" or "verificationist" who only believed in the reality of human sensation (Blackmore 1973). In this chapter, I give a positive characterization of Mach's physical elements, based upon my previous work (Banks 2002, 2003, 2004), and show how Mach was the first to develop the overall umbrella framework of realistic empiricism. The next chapter will focus on Mach's view of psychology and the mind–body problem.

Monism, not phenomenalism

As Mach often said, the leitmotiv of his thinking was not phenomenalism but *monism*, the search for a unified, metascientific world view encompassing sensations and physical phenomena under one roof. Such a view could not be physicalistic or materialistic in the traditional sense of the "physical," i.e., positing matter in motion and nothing else, for this would leave out the reality of phenomenal experience, or make it an insoluble puzzle, as Mach's contemporary Emil du Bois-Reymond had insisted. Idealistic monism was no better since it could not do justice to the reality of the mind-independent external world by making matter a mere bundle of sensations, or in positivistic terms reducing physics to a mere catalogue of observations.

Physicists had simply ignored the problem of sensation or relegated it to psychology. Psychologists had returned the favor by ignoring physics and relying upon introspection, with all of its attendant problems. Consequently, most of the progress on the unified science front was made by those hybrid nineteenth-century sense-physiologists and psycho-physicists like Müller, Helmholtz, Fechner, Hering, Stumpf, James, and others, who struggled with this problem seriously as part of their scientific research. Mach, for his part, sought a neutral position that was neither a materialistic monism *nor* an idealistic monism, but which would allow him to study sensations of color or sound in psychophysics alongside the physical phenomena of bodies and forces, without a fundamental change of perspective and without any fundamental mental/physical distinction.

Mach's elements

Mach was a realist about human experience. An element was only a sensation when one emphasized its functional relations to the human nervous system, mental images, and psychological laws of association;

otherwise the same elements could also comprise a physical object, functionally connected with other physical events in space and time. In itself, the element was neither mental nor physical, but a simple "neutral" occurrence:

> A color is a physical object as soon as we consider its dependence for instance upon its luminous source, upon other colors, upon temperature, upon spaces and so forth. When we consider, however, its dependence upon the retina ... it is a psychological object, a sensation. Not the subject matter but the direction of our investigation is different in the two domains ... it is only in their functional dependence that the elements are sensations. In another functional relation they are at the same time physical objects. (1886/1959, p. 16)

A sensation like a red patch thus counts as a *real* physical event, when it is described in physical terms and in relation to other physical elements, at least in an enhanced sense of the "physical," not in the narrower sense of "physics" which covers only mind-independent objects in space and time. Red is also what Mach called a "physical quality":

> Let us analyze mental experience into its constituents. First we find those that are called *sensations* insofar as they depend upon our bodies (the eyes being open, direction of ocular axis, normal condition and stimulation of the retina and so on), but are physical *qualities* insofar as they depend on other physical features (presence of the sun, tangible bodies and so on): the green of the park, the greyness and shapes of the town hall, the resistance of the ground I tread, the grazing contact with the cyclist, and so on. (1905/1976, pp. 15–16)

Although stated in the most sober terms, Mach's position is very bold and flies in the face of 300 years of considering sensation qualities like colors to be so-called secondary qualities with no existence outside the mind and certainly no physical reality. The method Mach used to arrive at this conclusion straight off seems to me to be the method of analysis, used in ancient mathematics, which he would have known. In synthesis, the method used by Euclid, propositions are built up from simple axioms to more complex truths. In analysis, we assume the problem solved, for example the doubling of the cube, and work backwards to analyze the conditions required for the solution. In effect, Mach has simply assumed that sensations *are* physical events and passed immediately to an analysis of the conditions under which that could be true and in which they could be investigated like physical objects and events.

Russell later credited Mach and James with a conceptual breakthrough, calling it a "service to philosophy," "that what is experienced may be a part of the physical world and often is so" (Russell 1914/1984, p. 31), and that "constituents of the physical world can be immediately present to me" (p. 22). Russell also uses the word "physical" here to mean an enhanced view of the physical world that includes sensation. After his own conversion to neutral monism, Russell was effusive in his praise of Mach, calling him the "inaugurator" of the whole neutral monist "movement" (1921, p. 16) and praising him highly to another Viennese, his former student Wittgenstein.[2]

For Mach, the important features of a sensation, which it shares with physical elements generally, are its concrete individual *existence*, its quality, and the fact that sensations are *events* that change rapidly and are quickly succeeded by others. Elements are not objects or states of objects, nor are they types, abstract concepts, or relations *between* objects, at least not fundamentally. All of these other terms are derivative and defined out of elements and causal–functional relations when we strip our experience down to the bare bones. Objects do of course exist, as indeed do interactions of objects, states, and the like, but they are one and all reducible to functionally connected elements (Mach 1886/1959, pp. 6–7, 29), some of which we observe and others of which we "add in thought" (pp. 15, 43–45) by completing the function of the object, substituting unobserved elements for the gaps in observation. Again I refer the reader to Mach's passage in the *Mechanics* about a vibrating rod (1883/1960, pp. 587–588) for his view that unobserved elements are indeed necessary for physics, but only once their continuity with experience is established. And of course once continuity is established there is no reason to consider the observable/ unobservable distinction as fundamental. What takes its place in Mach's monistic world view is a sort of continuous fabric of experience-reality consisting entirely of events and causal–functional relations both observed and unobserved (Mach 1905/1976, p. 361, Banks 2003, Introduction and chs. 7 and 9).

The physical reality of sensation

Human sensations were highly important to Mach, who was a major figure in the psychology of sensation and perception (Boring 1942, von Békésy

---

[2] Wittgenstein was gratified that Russell thought so highly of a countryman of his, but detested Mach's style of writing.

1960, Ratliff 1965, Banks 2001). He devised many ingenious experiments to measure their quality and intensity and their changes under a variety of conditions. Certainly he hoped such experiments would shed light upon the physiological basis for perception in the sense organs and the brain, but sensations had even a deeper meaning for him. For Mach, sensations were a peek, however limited, brief, and confused, at the nature of *reality*. For Mach, the blue of the sky that one sees is a real fact, as real as any in physics. Sensations are events that occur inside our brains while in certain highly complex physical states, but they do accurately manifest that reality as it truly is, and thus Mach believed sensation gave what he called a "real knowledge" of nature (Mach 1905/1976, p. 361).

Even Mach's admirers, like Paul Carus and Hans Kleinpeter, were surprised to find out just how realistic his attitude was toward human sensations. After conversing personally with Mach, they then warned against facile interpretations of the *Analysis of Sensations* that reduced Mach's position to the idealism of a Berkeley or Hume. In an 1893 article for the *Monist*, Carus reported on a meeting with Mach to clarify this issue:

> When several months ago I met Professor Mach in Prague ... he assented to my speaking of scientific terms as abstracts ... But when I proposed that the term "sensation" also was according to my terminology an abstract term presenting one feature of reality only and excluding other features, Professor Mach took exception to it saying that he understands by sensation reality itself. (Thiele 1978, p. 183)

Kleinpeter also gave an apt description of Mach's new realism, also in the pages of the *Monist*:

> Mach explained as the object of the naturalist's occupation "sense perceptions," a name which he chose under compulsion in lieu of a more subtle one, and which he later replaced by "elements"; for the sense in which he wished to have the word taken deviated to some extent from the customary usage. By the sound of the word we are inclined to put too much emphasis on the perceiving subject. But it was far from Mach's intention to emphasize this connection from only one point of view; on the contrary he saw at the time in sensation the material of the actual world. In this his fundamental views are essentially different from those of idealistic philosophers, Berkeley's among others. We may even call them realistic, but their realism is essentially different from so-called philosophical realism. (Kleinpeter 1906)

Mach's realistic view of sensation harkens back to a feature that has always been the animating spirit of empiricism, namely its focus on what is concretely and directly real in experience and existence. As complex and confused as sensations doubtless are, they put us in touch with nature in a direct way, even if what they show us are the complex internal states of our own brains since that too is part of nature, for Mach. I think it is this immediate and visceral sense of *existence* that experience conveys that accounts for the perennial appeal of empiricism to Mach and many other scientists and philosophers, the present author included.

Of course this very question, of sensations being real constituents of the *physical* world, was exactly the sticking point for other critics, like Vladimir I. Lenin, who wrote a famous polemic against Mach, Avenarius, and their Russian followers such as Bogdanov: *Materialism and Empirio-Criticism* (1908/1952). Lenin's book could be considered the beginning of those positivistic and idealistic readings by Schlick, Popper, Blackmore, and others that plagued Mach's work and legacy. As one might expect, the book reads more like a political pamphlet. Lenin asked: how could the nature of the physical world of body, of atoms, of force, ever be found in sensory qualities? Qualities are *mental*; there are no qualities in *matter*, he insists. The only possible conclusion, given Lenin's stark dualism, was, predictably, that Mach was just an idealist after all, seeking to reduce real mind-independent matter to human sensations:

> Mach and Avenarius *secretly* smuggle in materialism by means of the word "element," which *supposedly* frees their theory of the one-sidedness of subjective idealism, *supposedly* permits the assumption that the mental is dependent on the retina, nerves and so forth and the assumption that the physical is independent of the human organism. In fact, of course, the trick with the word "element" is a wretched sophistry, for a materialist who reads Mach and Avenarius will immediately ask: what are the "elements"? Either the "element" is a *sensation*, as all empirio-criticists, Mach, Avenarius, Petzoldt, etc. maintain—in which case your philosophy, gentlemen, is *idealism*, vainly seeking to hide the nakedness of its solipsism under the cloak of a more "objective" terminology; or the element is not a sensation—in which case *absolutely no thought whatever* is attached to the "new" term; it is merely an empty bauble. (1908/1952, pp. 48, 49)

Lenin is captive to the very dualism that Mach and Avenarius so strenuously denied, but it is remarkable how many of his subsequent critics simply fell into line behind this plainly unfair reading. It seems to be too hard for Lenin or other interpreters to imagine, not even for the sake of argument, that a sensation is a real natural occurrence on par with events in physics, as Mach insisted, an insight which James adopted and which Russell considered "a service to philosophy." Of course the same misreadings greeted both James's and Russell's own neutral monist works (see in particular Lovejoy 1930 and Stace 1946/1999) and some would still argue today that sensation qualities like red or blue must be mental and cannot be neutral (see Bostock 2012).

It has been pointed out to me that Mach is hard to read and understand in the original, which is true. Mach wrote in a pithy, oracular, Old-World style in which it was considered bad form to say in many words what could be said in few. He left much to the intelligence and discernment of his readers. Moreover, his expertise extended to many different departments of inquiry which he assumes his reader knows. When he uses physical examples he is assuming his reader understands physics and can read meaning into his examples. Consider one of his favorite phrases: phenomenological physics. Mach's frequent reference to so-called phenomenological or mechanism-independent laws of physics, such as the excluded perpetual motion principle, or the Carnot–Clausius version of the second law of thermodynamics, or Kirchhoff's law of black-body radiation, derived from the second law, are not understood by those philosophers who think "phenomenological" means directly observable to the senses, an unforgivably superficial corruption of a deep and beautiful physical truth about the laws of nature. Philosophers may not be aware that there are laws of physics, such as laws of conservation, the principle of least action, the principles of thermodynamics, or other postulates of experience that do not invoke underlying realizing mechanisms or realistic causation, and that this is not a philosophical view, but a fact. These laws and postulates hold of natural events whether they are directly observed or not and the distinction is between two different types of physical law; it has nothing to do with observation. In short, at risk of being overly prolix myself, Mach is not easy or readily accessible without effort, and it is thus much easier to follow and repeat secondary popular accounts with a few collected quotations, usually taken out of context, instead of reading and pondering Mach himself.

Lenin is quite correct about one thing. We are not used to thinking of the standard "physical" world of objects as constructed out of evanescent *events*. And the idea that these concrete physical events will possess concrete *qualities* does indeed make one think solely of human sensations with the qualities of color or sound, as if an idealistic reconstruction of matter into sensations were being urged. The contemporary reader might well ask, with Lenin: Why is this *not* an attempt to "mentalize" reality? What else *could* these physical qualities be, if not sensations, or proto-sensations? How could physical events in mind-independent nature exhibit their own concrete *qualities*? Mach is attempting to revisit the basic division laid down during the scientific revolution by eminent scientists like Descartes, Galileo, Locke, Newton, and others, in their own battle against the natural "qualities" of Scholastic philosophy and earlier philosophers like Duns Scotus and Nicole Oresme in the Middle Ages. In the scientific revolution (leaving out dissenters like Leibniz and others) qualities of all sorts were permanently banished to the subjective realm of human sensation. Newton even called them "phantasms" and denied qualities any physical reality at all (in his "Rules for Reasoning in Philosophy" in the *Principia*). As Galileo and Locke had insisted, Newton also held that the *quantitative* primary qualities of matter and motion (extension, bulk, motion, number, size, shape, position, etc.) go all the way down and are simply all there is to nature, as if nature itself were a kind of erector set or a diagram in a geometry book. And given the success of natural science, in the seventeenth-century pattern of quality-less matter and motion, why should we change our view of the physical simply to find room for our sensations in our physical world view, when it makes much more sense to eliminate them?

This is certainly a powerful objection and places the burden of proof squarely on Mach and others to show exactly what the adoption of elements and their qualities would possibly add to science. We see them, but how do we know that they *are* real and not illusions like the rainbow? How could my private sensations or yours be made accessible to objective physical inquiry or measurement? This appears to me to be the main obstacle to the understanding and adoption of Mach's proposal to consider sensations—colors and all—as real physical events, again in that extended sense of "physical" that he intended to introduce.

The umbrella framework: elements and functions

First it helps to see exactly how Mach sought to change physics. There is a remarkable consistency in Mach's views on physics over the many decades from his *Conservation of Energy* (1872/1910) all the way through to the late *Knowledge and Error* (1905/1976). He ended all of these works by laying out a programmatic design for future science based upon his function-and-element methodology, saying that the job of natural science is to determine the natural variations in the elements expressed by those functions, expressing the reciprocal dependencies of the individual elements on one another, in objects, systems of objects, and human minds:

$$F(\beta,\gamma,\delta\ldots\omega) = \alpha$$
$$G(\alpha,\gamma,\delta\ldots\omega) = \beta$$
$$H(\alpha,\beta,\delta\ldots\omega) = \gamma$$
$$\psi(\alpha,\beta,\gamma,\delta,\varepsilon\ldots\omega) = 0$$

This is what I will call an *umbrella theory*, a theory schema for the construction of specific empiricist scientific theories on this model of the austere "element-and-function" methodology. An umbrella theory gives a sort of pattern to follow in constructing empirical theories of the world but it does not pronounce on the specific details. It shows what this class of theories will look like in their general features but it cannot make specific predictions; that is left to the theories themselves. So one might look at special relativity in Einstein's 1905 paper, or Heisenberg's seminal 1925 paper on quantum mechanics, as exemplars of a theory schema like the one Mach is proposing. Actually what appears to have happened is Einstein drew upon Mach's model and then Heisenberg used Einstein's theory as a schema for constructing his own theory of the atom (Heisenberg 1986).

What is very unclear at this stage is how to translate standard physics of extended bodies and motions into an event-and-function-based framework like this. Physicists do talk about events and sometimes even say that they are fundamental, but they do not actually theorize in these terms. Physical quantities are traditionally built up in dimensions of mass, length, and time and refer to the motion of extended objects like particles and fields in space and time. Various interactions between these more basic objects are what is meant by a physical event, so events are not fundamental to physics after all. There is already a kind of prior intuitive framework for physics in place before there is any talk about physical events. Laws like Newton's second law, the conservation of energy, and least action, express higher-order abstract relations of a system of particles and forces. But these do not appear to be pure causal–functional relations couched in "variations" and "counter variations" of "elementary events" of the sort that Mach posited as fundamental. The most fundamental things in physics still seem to be particulate matter and motion and fields in space and time.

Hence, I think we should regard Mach's element-and-function methodology as an attempt to actually *change* physical science rather than rationally reconstruct it as it exists (on this point, see Feyerabend 1970, 1984). Mach was never bashful about criticizing physics, with excellent results: his searching critiques of concepts in Newtonian physics (space, time, mass, energy) and nineteenth-century thermodynamics showed his refusal to fall into line and simply recite statements of principle verbatim. He was always doing a fresh historico-critical investigation of fundamental concepts, which were the real subjects of all of his books. To Mach, science was a system of open critical inquiry, never a set of stock textbook principles and set problems to be learned by rote. Mach was also aware of important gaps in the conceptual structure of physics which he believed only an historical-philosophical analysis of physics could reveal. Long before Thomas Kuhn and the twentieth-century reaction against logical positivism led philosophers to the historico-critical method again, Mach realized that these anomalies in physics are the beginnings of new science and should not be ignored but rather brought to the forefront of inquiry.

Mach's main target for criticism was the "mechanical philosophy" of the seventeenth century, its metaphysical background, and its basic concepts. As he often said (for example 1883/1960, pp. 596–597), mechanics is not the fundamental science, but deals with objects and motions readily accessible to the senses. It comes first, historically and epistemically, but it need not go deepest. Rather the laws of mechanics and basic dimensions of mass, space, and time serve as analogies for the discovery of still deeper natural laws which need not be grounded in mechanical systems and which instead may be considered "phenomenological" principles, or "postulates of experience," grounded in natural events themselves and lacking nothing further (1883/1960, p. 599). Here is a typical passage from his 1896 *Principles of the Theory of Heat* expressing what he meant by a "phenomenological" or comparative physics, as opposed to mechanical physics:

> Physical processes present numerous analogies with purely mechanical ones. Differences of temperature and electric differences equilibrate themselves in a similar way to the differences of the position of masses. Laws which correspond to the Newtonian principle of reaction, to the law of conservation of the center of gravity, to the conservation of the quantity of motion, the principle of least action, and so on, may be set up in all physical domains. These analogies may be made to rest upon the assumption which the physicist is fond of making, namely, that all physical processes are in reality mechanical. But I have long been of the opinion that we can discover general phenomenological laws under which the mechanical ones are to be classed as special cases. Mechanics is not to serve for the explanation of these phenomenological laws but as a model in form and as an indicator in searching for them. The chief value of mechanics seems to me to lie in this. (Mach 1896/1986, pp 328–329)

Mach's powerful criticisms of mechanics convinced many of his contemporaries and those on their way up such as the young Einstein:

It was Ernst Mach, in his history of mechanics, who shook this dogmatic
faith (i.e., in mechanics as the secure basis of physics); this book, in this
respect, exercised a deep influence on me as a student. (Einstein 1949, p. 7,
and for the possible influence of Mach's non-mechanical principle-driven
"phenomenological physics" on Einstein, see also pp. 53, 57; also
Feyerabend 1984, Banks 2003, p. 188)

Mach believed that in order to lay these issues bare, the preliminary stage is
to get present physical theories into the "bare-bones" phenomenological
form of the element–function theory schema he had outlined. This gives
the real content of the physics shorn of the imagery of the mechanical
world view. He gives an example of how to eliminate extraneous "fluids
and media" from electrodynamics, such as electric and magnetic flux lines,
replacing them with measured electrical potential and level values (1883/
1960, p. 598). He even boldly imagined the eventual elimination of the
concepts of space, time, and extended particles or fields from fundamental
physics (see Banks 2003, pp. 13–14, 29, 34), which he thought would deal a
death-blow to the mechanical philosophy, replacing it with his deeper
theory of elements. So Mach's theory schema functions first as an engine
of analysis for streamlining existing physical theory (see the conclusion of
his 1872) to emphasize its purely phenomenological character and downplay and finally eliminate mechanistic picture-thinking and spatio-
temporal visualization processes. In this way, the accidental historical path
of development of physical concepts could be purged from the science,
along with any historically conditioned a priori assumptions and any
additional sensory, visual content, or realizing mechanisms that did not
play an indispensable role in the original theory.

As a working physicist, Mach was interested in the nature of energy and
energy transitions across the various departments of physics. The language
of potential functions and forces had been generalized by Mach's time
beyond its roots in mechanics, in the work of Clairaut, and extended to
electricity and magnetism by Thomson and Maxwell, as well as thermodynamics by Duhem. For the first time a kind of universal physics seemed
to be in the offing, unifying the different departments of inquiry under one
set of laws, as J. R. Mayer had planned. These developments suggested to
many physicists of the time, including the young Max Planck, the idea of a
"universal dynamics" based on general laws of energy transfer and Hamil-
tonian and Lagrangian least action principles (see Hiebert 1968). Where
"the energy" had originally been just one integral of Newton's $F = ma$
with respect to space (the other, time-integral, being momentum), by
Mach's era, the conservation of energy, discovered by Mayer and Joule,

had indicated a much more fundamental role for the concept of energy than the original application to mechanical systems, the lifting and falling of weights, Leibniz's law of vis viva, and the impossibility of perpetual motion. It may perhaps be clear that mechanical work and heat are the same, if heat is just the vibration of molecules underneath, but it is quite a leap to extend the principle to electromagnetism, where totally different things like charges and fields of force obey the principle as well, for example in Lenz's law.[3] And it is certainly a huge leap to extend the principle to any and all natural events operating by whatever mechanisms, or even no mechanisms whatever, as Mach does at the end of his (1872) book.

Responding to this current of thought, as I have shown (Banks 2003, chs. 10–11), the elements, which Mach had encountered in psychophysics and J. F. Herbart's psychology and metaphysics, were actually imported into Mach's physics as concrete changes of a certain quantity of energy (potential) from one "level value" (potential level) to another, across the different branches of science. The theory of elements, if they are interpreted as energy changes, or what Mach simply calls "differences," could thus be broader than mechanics, or any mechanical view of nature (Mach 1905/1976, pp. 357–358). These energy quantities and their changes formed the empirical backbone of scientific theories, just as phenomenological laws or postulates of experience formed the basis of abstract functional connections among the elements, or the law governing potential differences. This kind of project for a universal science of energy change and abstract function was the theme of Mach's short 1872 book entitled *The History and Root of the Principle of the Conservation of Energy.*

As historians of science are well aware (see Hiebert 1968, 1973, Banks 2003), Mach's view of the energy principle had its idiosyncrasies. For example, Mach implicitly spoke of an energy quantity or potential amount (E) always at a certain level value or potential level (l). Strictly speaking, this term (E/l) is dimensionally equivalent to the entropy (S), as Mach realized, not the energy term (E) alone. According to Mach, however, *both* terms were needed to describe the actual potential to *do* work, which was the historical basis for the energy concept, not merely the quantity term (E) but also the level value (l): see Figure 0.3. Using an "energetic analogy," originally due to Zeuner, Mach observed that the tendency of natural forces was to transfer quantities of energy from higher to lower potential

---

[3] Lenz's law predicts that as a magnet passes through a coil, or vice versa, the induced EMF in the coil resists the motion in the opposite direction. If it did not, the magnet would accelerate and energy would not be conserved in this combined system.

levels, the way masses fall from higher to lower heights, quantities of heat flow from higher to lower temperatures, or current flows from higher to lower voltages. Others had noticed these cross-physical analogies of course, Rankine, Thomson, and Maxwell among them.

Mach's view, however, should *not* be assimilated to the later energetics movement of Georg Helm and Wilhelm Ostwald, who also asserted a variety of analogies, many of them flawed (see Hiebert 1973 on this point of letting Mach stand apart). For Mach, energy was *not* a "universal stuff" for example, flowing from place to place like a fluid as Ostwald seemed to think. There was no all-purpose energy *an sich* beneath appearances waiting to manifest one way or another, there were only the individual manifestations of concrete energy changes, the "differences," manifested in transitional events from one form to another and governed by abstract functions, like the potential. Only these direct manifestations of energy *differences* in events could be considered physically real. Again, we can see Mach stripping away as much as possible even from the energetic theory. The transitional *events* of energy transfer are fundamental, not the potentials themselves. That's a very important point. Thus, even the "quantity" and "level" of potential, seized upon by the energeticists, are abstractions grounded in aspects of the concrete events we can pick out (see Banks 2002, 2003 for details).

During their famous debate in the early twentieth century (for which see Hiebert 1968, Blackmore 1992) Mach was accused by Planck of misunderstanding the basic difference between energy and entropy, between the first and second laws, and especially between reversible (Carnot engines) and irreversible processes (such as gas expansion). As I have shown (Banks 2003, ch. 13) Planck's criticisms were mistaken on this score. There are no confusions in Mach's works about these concepts, or the differences between them, *when* we allow for the basic fact that Mach understood by energy (E/l) an energy amount E plus level value l combination, which is dimensionally equivalent to the entropy (Q/T) not the energy term alone. This is admittedly a non-standard usage, based on Mach's study of the history of the concepts of energy and entropy, which probably did lead Planck and others to think Mach had confused these terms when he was simply defining them his own way.

Mach was perfectly capable of understanding the expansion of a gas as a case where a quantity of energy is *not* lost or transferred and yet something happens. It does not "transform," as he says other energies do when they fall in level. Yet the potential/level language can still be used to describe the change, as we still do today, as a metaphorical way of thinking of the heat as remaining the same in quantity but falling strictly in its level value, which results in an entropy increase. By letting the gas expand we have lost some metaphorical work it "could have done" and it will thus take energy put into the system to restore it to this initial state, indicating an irreversible natural process. The description of the expansion of a gas as "a 'fall' in its capacity to do work" is the purest metaphor yet it is grounded in the real differences of initial and final states observed in the events which it describes. Hence also Mach's reverse complaint directed at Planck and others that physics had become a "church" where questioning and redefinition of basic concepts was no longer allowed. Mach wanted to describe energy as the capacity to do work between levels and not as the simple term for the quantity of energy, E. Rigid attitudes about physics were about to change suddenly around the turn of the century, and I believe we should always think about what happened next when looking at these debates, giving ourselves the benefit of hindsight which the principals did not have.

So, as to the physical definition of a Machian element, we are down to the concept of concrete events manifesting energy quantity plus level changes. We understand these transitional events, of whatever kind, across physics, by treating them via a concept, the energy, which developed historically by analogy with the ability to perform mechanical work by raising a weight. It is not just the quantity (E) that matters, but its potential level too (E/l), and what matter most are the observable, transitional events. All else besides the transitional events or "differences" themselves, even the metaphorical quantity and level terms of energetics, belongs to the surrounding superstructure of functional connections, as the theory is further stripped down to basics.

This "phenomenological" view, based on thermodynamics, can never be a *complete* view of the laws of physics of course, as Mach himself realized, and which was pointed out sharply during the famous energetics debate at the 1895 Lübeck *Naturforscherversammlung* between Boltzmann and Wilhelm Ostwald, and afterward in a post mortem, in which Planck and Boltzmann participated to put the final nail in the coffin of Energetics (see Deltete 1999). As Max Jammer (1963) showed in a review of Hiebert, it is often unclear how to define these analogues of "potential" and "potential level values," at least in conventional physical terms. And, as Deltete shows in detail, energetics in the hands of Ostwald and Helm postulated some strange things like "volume energy" which were not always state functions, meaning you could not follow these quantities in and out of the thermodynamic system. Heat Q is not a state function, for example, because it can flow into and out of a gas by conduction, but it can also be developed by doing work on the gas or drawn off by letting the gas do work on a piston. Shown a certain quantity Q, we literally don't know where it came from or where it went unless we also know the work done on, or by, the gas, or other factors which can give us the required information. Entropy, or heat divided by temperature on the absolute Kelvin scale, for example, *is* a state function and describes the system independent of the path taken, either by heating the gas or doing work on it, since both change the entropy of the system regardless of what is doing it.

But even if these energetic analogies could have been articulated more skillfully, the law of "equalization of energy differences" (Mach 1886/ 1959, pp. 351–352, Banks 2003, ch. 11) is still much too thin to predict the exact nature of the change, the *path* or direction of motion for example, or even whether a change will occur or not, as Mach of course realized. He himself pointed out that the law of conservation of energy could predict that a body will "hang in the air indefinitely" (Banks 2003, p. 199). Simple energy data are not enough to predict an energy change (Mach 1896/1986, p. 313, Banks 2003, pp. 168–169); for that, we need a basic law of motion as well, like Newton's second law or the principle of least action, or Schrödinger's equation. Mach suspected there would be a deeper "law of energy changes" or "law of differences" as well as a law for explaining the "arising of potential differences" but he did not know what it was. We thus have physical elements with Mach, but he still lacked the basic functional laws that govern energy changes. So, while many of the attacks on energetics and Mach's views, by Planck (see Blackmore 1992) were *ad hominem*s, and based on misunderstandings, Planck was ultimately correct to point out that energetics never did offer more than an outline of a view of physics, more like a philosophical ideal than a concrete program.

There is an important epilogue to this story, which should be mentioned even if it cannot be fully developed, and even though I am committing a blatant anachronism by telling it. The phenomenological and mechanism-independent science of energies and happenings that Mach and the energeticists had hoped for eventually turned out to be the quantum theory, as developed by Bohr, Heisenberg, Schrödinger, Dirac, and their collaborators in the first quarter of the twentieth century. The quantum theory is a fundamental, postulate-based science which describes for example the amount of energy manifested in an event, such as when an electron falls in toward the nucleus, in quantized jumps, and emits a quantized photon of a given frequency. And more importantly it also describes the *probability* of such a jump occurring, or the important law governing the energy transitions, even if the law is not deterministic as Mach and others had hoped it would be.

How was this finally discovered? The immediate physical problem was to explain the intensity of radiation in the bright line emission spectra of atoms. The frequency of the emitted photons was proportional to their energy and thus to the difference of energy levels fallen through by the electron toward the nucleus, by the Planck–Einstein relation $E = h\nu$, where the orbits were further articulated by Bohr, using the quantum condition $nh = \oint p \, dq$. The remaining problem, which was first solved by Heisenberg, and then again by Schrödinger, was to determine the intensity of the spectral lines for any given frequency. This corresponds to the *number* of photons emitted at each frequency for statistically large numbers of jumps and emissions, or, in the single case, to the probability of an atom emitting one photon of a given frequency. The probability of an energy transition is, in turn, expressed as the square of an "amplitude" of a wave given by Schrödinger's wave equation or the square of a "transition coefficient" using Heisenberg's matrix methods. This result was inferred by analogy with classical physics. Given that we are looking for the analogue of an energy (a number of emitted photons), and given the fact that the classical energy of a wave is the square of its amplitude, we assume *another* amplitude of some other "hidden" oscillatory wave motion in the atom of which the square will give us the intensity of the emitted photons. This oscillatory wave motion is a fiction, in the sense that the waves cannot be observed directly, and the wave is expressly *not* the real three-dimensional motion of an electron, as Schrödinger once believed. Even in cases where we seem able to visualize three-dimensionally, the oscillation of the wave function is still only an abstract "motion" in configuration space, an infinite-dimensional Hilbert space, which only takes real form when a measurement is made on the system.

Now as far as anyone knows, this is as deep as the explanation of energy transitions goes, or will ever go, and the situation does seem rather Machian. The observable quantities are the energy differences, or what Mach would have called "elements," and their transition probabilities are indeed given by an abstract wave function with no underlying mechanical system or model behind it, only an abstract law or function of possible observations that could be made on a system and lead to events. As Richard Feynman once put it, no "machinery" has ever been found to explain the wave function and no one expects there to be any. Hence the quantum theory does indeed give the kind of "stripped down" empirical theory Mach would have championed, and not just in its early days, but even in its final development too in the form of postulates of experience governing natural events, not mechanisms or mechanistic causation. This, in my view, makes the struggle with Planck all the more poignant. Planck, who introduced the quantum into physics in his theory of black-body radiation, was defending the mechanical philosophy and Mach, the skeptic of atoms and classical physics, lost the argument in his own time, but was defending a view of theories that turned out to be very appropriate to the quantum era.

Reality of forces

Manifested potential differences are forces for the physicist. Clearly, forces cum potential differences were very concretely real for Mach: pushes, pulls, shocks, stresses, tension, pressure manifesting in a whole variety of colorful events of different types emanating from different natural sources, masses, inertial forces, electromagnetic forces. It is truly ironic that Max Jammer (1999, pp. 211–229) and others lump Mach in with the program to *eliminate* force from mechanics, for example, by Hertz and Kirchhoff, both of whom regarded forces as occult and worthy of elimination in terms of the spatial accelerations of masses in motion. This is a completely false comparison (see Banks 2003, pp. 194–196, Preston 2008). Mach is more like Leibniz or Boscovich in affirming the basic reality of force, or more precisely those individual events in which forces are manifested, denying the ultimate reality of bodies in motion, which is the complete opposite of the view usually attributed to him in physics. Truly, this is yet another example where Mach has been ill served by his interpreters, who do not even take the trouble to read him on this point. For example, Mach wrote apropos of Hertz's *Mechanics*:

> To characterize forces as being frequently empty-running wheels as being frequently not demonstrable to the senses can scarcely be permitted. In any event "forces" are decidedly in the advantage on this score, as compared with "hidden masses" and "hidden motions." In the case of a piece of iron lying at rest on a table, both the forces in equilibrium, the weight of the iron and the elasticity of the table are readily demonstrable. (1883/1960, p. 319)

Of course, the further hypothesis that these concrete energy changes manifested by forces must manifest concrete *qualities* of their own goes well beyond physics and is an added metaphysical hypothesis, i.e., that sensation qualities too *are* physical events and perhaps also manifest forces of some kind (an idea he may have got from J. F. Herbart, as we will see below). Mach, too, was doing metaphysics, albeit of a very austere empiricist sort that he did not even perceive as a metaphysical view of nature, but rather, perhaps, simply as natural philosophy.

The atomism episode: *"Hab' Sie schon Eins gesehen?"*

Mach's vociferous opposition to atoms cemented his reputation as an ultra-positivist who did not believe in unobservables. This is yet another area where historical stereotypes have obscured the issue and I think it appropriate to add a few remarks about this here. For Mach, it is true that the basic happenings in nature are not little atomic bodies but force-like elements which make up bodies and systems in functions. In fact, Mach's elements were intended to go much deeper than what he saw as the psychologically comfortable but misleading visualizations of little billiard balls, colliding with one another and attached to each other by little springs, or other such mechanisms.

Mach's skepticism of atoms (first announced vigorously in his 1872/1910) was actually prompted, as he says, by the problem of the lines of the emission spectra of atoms, for there were too many of them to be accounted for by vibrations of little billiard balls in three-dimensional space, a legitimate scientific issue at the time. The idea of an atom with internal structure never seemed to occur to him, nor did atoms which were not *Dinge an sich*, but just very small objects continuous with experience, at least not until this was demonstrated to him. In this respect, for most of his career anyway, Mach's view of atoms was very nineteenth-century indeed, tinged with Kantian metaphysics. In another respect, Mach was looking ahead, as I have written recently (Banks 2012), since one could see his tabular view of atomic vibration in the 1872 booklet as an anticipation of Heisenberg's matrix atom, which at least one author has suggested (Weinberg 1937).

As Paul Feyerabend pointed out in his 1984 reappraisal of Mach's philosophy, elements are more basic than atoms, bearing in mind Mach's aforementioned eliminationist program for spatio-temporal notions and objects. Moreover, Mach saw atoms as little sensory objects and even as Kantian "metaphysical monsters," substances unobservable even in principle:

> The vague image which we have of a given permanent complex, being an image which does not perceptibly change when one or another of the component parts is taken away seems to be something which exists in itself.

Inasmuch as it is possible to take away singly every constituent part without destroying the capacity of the image to stand for the totality and to be recognized again, it is imagined that it is possible to subtract all the parts and to have something still remaining. Thus naturally arises the philosophical notion, at first impressive, but subsequently recognized as monstrous of a "thing in itself," different from its "appearance" and "unknowable." (Mach 1886/1959, p. 6)

I think hardly *anyone* party to the atomic debate in the nineteenth century believed physical atoms to be Kantian *Dinge an sich*, and it is certainly not the way we think of atoms today. His other question about how to visualize the processes interior to atoms is, however, a legitimate issue, and does invite skepticism about space-time visualizations of atomic processes such as the emission and absorption of radiation, which would be center stage in the early twentieth century during the development of quantum theory by Bohr and Heisenberg.

So the question about who was ultimately right about atoms in those late nineteenth- and early twentieth-century debates is a hopeless anachronism. Mach, whose view of atoms was clearly a "metaphysical monstrosity" of his own invention, was eventually convinced by Stefan Meyer of the reality of little objects of some sort in 1903 (Blackmore 1992, p. 151) and then, perhaps, he immediately turned his attention to the interiors of those nuclei, finding only more elements. I do not think this conversion to atomism thus led to any major revision of his views about the element-and-function methodology he had adopted; however, he considered his elements more fundamental than atoms, representing perhaps the energy-transition events in their innards. In that sense the comparison with Heisenberg's matrix mechanics may be more apropos as the true continuation of the Machian program, but this would require further study.[4]

Elements as "forceful qualities"

Force, as it turns out, is the linking concept between the way Mach characterizes physical elements, i.e., physical events which are the direct manifestations of potential differences being traversed, and the concrete empirical manifestations of individual qualities in those events, if quality is

---

[4] There are certainly connections here. Perhaps Heisenberg also knew of Mach through Wolfgang Pauli, whose father was an old friend of Mach himself. Mach was Pauli's godfather. Heisenberg also speaks of philosophical discussions with friends prior to his first paper on matrix mechanics. All of this would be very interesting to fully flesh out.

the actual concrete manifestation of natural forces, when we treat the events as individuals and not as more abstract types or classes of events, as is often the case in physics itself. For lack of an appropriate philosophical term, I will call Mach's elements "forceful qualities," to express the idea that the elements are events with concrete natural qualities like our sensations, but also to express the claim that the qualities of these elements are actual individual manifestations of natural forces (Banks 2003, Introduction, chs. 3, 7, 9). This basic idea that a quality directly manifests force or power has been rediscovered in the notion of "powerful qualities" by Martin (1993) and Heil (2003, 2005).

As I have shown at great length, Mach arrived at this view of elements cum "forceful qualities" early in his career, around 1866, probably through his study of the psychophysics of G. T. Fechner and especially the *Psychologie als Wissenschaft* of the German philosopher J. F. Herbart (see Herbart 1964, Banks 2003, chs. 2 and 3, and 2005). This is the underlying reason why the elements fall naturally into functions and complexes, in which their reciprocal changes in intensity can be related to one another. It is also how phenomenological functions find their grounding in real relations of force between the elements, and not mere thought economy, which really has to do with higher-order groupings of patterns (see Banks 2004). Elements *are* manifested forces in events. In physical events, these "forceful qualities" will not be blue or red like our sensations, which are complexly structured natural events in our brains. But physical elements do have concrete qualities of their own, through which they manifest their powers, and for Mach these can be compared to the qualities we are familiar with in our sensations, for that is our closest analogy at the moment.

As I have shown (Banks 2002; 2003, ch. 3), Mach probably adapted the notion of a "forceful quality" from the German philosopher J. F. Herbart, a major influence on Mach and other important philosopher-scientists in Germany like Bernhard Riemann and Hermann Grassmann (see Banks 2005, 2013c). Herbart was well respected and widely read, but he had the misfortune to be a realistic, mathematical philosopher during the heyday of German Idealism, which he strongly opposed. Herbart was also perhaps the first person to attempt a genuine mathematical psychology, *Psychologie als Wissenschaft*, which dealt with qualities of sensation as forceful entities amenable to scientific mathematical treatment. For Herbart, the *type* of a quality, such as the hue of a color, or the pitch of a sound, was analogous to the direction of a force, and qualities were further directly identified with forces by possessing a certain power or intensity. Herbart represented colors, for example, as having qualitative directions within a color space, a color triangle with red, yellow, and blue on the corners serving as the pure directions of the manifold, the way that right and left or up and down give directions in geometric space. Pitches were likewise arranged like right and left in a tone space. These qualities opposed and resisted one another dynamically according to natural laws; they did not just "appear" inertly in our experience like sense data, or epiphenomenal qualia. Rather, experience was the result of "psychical energies" in the nervous system, striving to inhibit each other and reaching a dynamical equilibrium of forces which is the basis of consciousness in Herbart's system, a so-called "apperceptive mass" in which all sensations were understood chiefly as qualitative expressions of psychical forces straining against each other.

This Herbartian comparison of sensory qualities with natural forces was not new; the idea can be found in Aristotle, in Medievals like Scotus and Oresme, and in Leibniz and Kant. What *was* new was Herbart's attempt to give a serious, quantitative mechanics of sensory qualities, a *mathesis intensorum* or "science of intensities," which would do for the qualitative data of psychology what Newton had done for physical forces. With the direction given by the quality of the sensation, its vivacity gave a measure of the intensity of the force that it conveyed to others of the appropriate opposing type. A more intense sensation would thus overcome a less intense one and press it, almost like a spring, below the limin of consciousness, where it could later be awakened, if the barrier were removed. Sensations of independent types, meanwhile, simply fused to a complex perception, without pressing against one another, the way orthogonally situated forces would do. This mathematical treatment of forceful qualities by Herbart seems to have made a deep impression on the young Mach when he developed his theory of elements in the 1860s (Banks 2003, chs. 3 and 4).

The "laws" or "functions" Mach speaks of are not therefore mere laws of mental association, or mere abstract mathematical functions which could conceivably relate anything to anything. Functions are grounded in the interplay of real forces, which was also the way Herbart understood them. Herbart had nuclei too, but these were little more than functions of their qualities. The causal–functional connections, relations, and variations Mach often speaks of should be understood as grounded in the underlying behavior of his force-elements. Mach was famously skeptical of causal relations in science, calling them animistic or sometimes an occult holdover from more primitive times. What I think he meant by this, however, is not that expressions of force or power are to be replaced by purely abstract mathematical functions, but rather that these kinds of power manifestations in events and relations of force between them should not be grounded in hidden mechanical mechanisms or visual models, but should be taken as real on the face of it, without a need for any further grounding in anything but the qualities and variations of the elements themselves.

A point I made in my article on Mach's economy of thought (Banks 2004) was the following. "First order" causal–functional relations are grounded in the real causal behavior of the elements and their qualities as they exert forces on each other. They are really there and not subject to thought economy at all. But because there will be many different kinds of first-order functional connections grounded in these qualities and their changes, including polyadic and even possibly macro-causal relations between islets or clusters of elements (see Mach 1872/1910, pp. 70–71), a second-order sorting of these patterns is necessary for the purposes of economy or codification of many phenomena under a few laws and principles.

Some causal–functional relations will be empirical laws like Snell's laws unifying a set of refractive indices and angles of refraction, or Kepler's laws unifying planetary positions and velocities into elliptical orbits; others will be more abstract theoretical principles governing events such as Fermat's principle of least time, the first and second laws of thermodynamics, or the law of least action in mechanics. As I indicated, "thought economy" dictates a second-order deductive arrangement of the available first-order patterns, not necessarily unique, where the more specific empirical regularities are brought under the general theoretical, abstract regularities of experience, such as deducing Snell's law from Fermat's or Newton's second law from the law of least action in mechanics. Mach does not actually call this an "explanation," as Carl Hempel would later maintain of nomological-deductive relationships, rather he calls it "the reduction of many uncommon unintelligibilities to a few common unintelligibilities" (Mach 1872/1910, pp. 55–56). No one really knows why the law of least action holds, but if it does, other patterns become intelligible on the basis of assuming it. The structure of physical science at the second order is thus dictated in some ways by economical need, and also by the historical paths of discovery, but these second-order factors cannot and should not replace the first-order objective functional patterns in the phenomena of individual events and their individual causal–functional relations, which are not up to us to choose (see Banks 2004).

Machian physics: relationism and anti-mechanism

Mach the physicist actually looked to Herbart's forceful qualities as inspiration for a fresh view of physics, based on physical elements, compatible with sensations, rather than the matter and motion of the mechanical philosophy (see Banks 2003, chs. 2 and 3). This intended, wide-ranging, reform of physics was part of an extensive philosophical and what Mach called a "sense physiological" criticism of nineteenth-century classical mechanics in his books on the *Conservation of Energy*, the *Science of Mechanics*, and the *Principles of the Theory of Heat*.

One form these thoughts took was of course Mach's well-known relationism about space and time. Mach believed that only relative motions could be observed, even in cases of accelerated motion. The actual underlying physical events would thus be the same whichever state of motion the observer adopted, since motion relative to a natural phenomenon is, for Mach, just a redescription of an underlying invariant physical event which is the same in either case. In particular, Mach believed that inertial forces and gravitational forces could be seen as the same, perhaps the effects of observing events occurring in an underlying invariant "inertio-gravitational field" which could appear different (either as gravitational force or inertial forces) when redescribed from differently accelerated reference frames, a field-view Mach actually suggested already in the 1883 first edition of his *Mechanics*:

> It might be, indeed, that the isolated bodies A,B,C ... play merely a collateral role in the determination of the body K, and that this motion is determined by a medium in which K exists. In such a case we should have to substitute this medium for Newton's absolute space. Newton certainly did not entertain this idea. Moreover it is easily demonstrable that the atmosphere is not this motion-determinative medium. We should therefore have to picture to ourselves some other medium filling, say, all space, with respect to the constitution of which and its kinetic relations to the bodies placed in it we have at present no adequate knowledge ... Although, practically, and at the present nothing is to be accomplished with this conception, we might still hope to learn more in the future about this hypothetical medium; and from the point of view of science it would be in every respect a more valuable acquisition than the forlorn idea of absolute space. When we reflect that we cannot abolish the isolated bodies A,B,C ... that is, cannot determine by experiment whether the part they play is fundamental or collateral, that hitherto they have been the sole and only competent means of the orientation of motions and of the description of mechanical facts, it will be found expedient provisionally to regard all motions as determined by these bodies. (Mach 1883/1960, pp. 282–283)

Mach hoped that inertial forces could somehow be due to the presence of distant masses, either directly or through a combined inertio-gravitational field, possibly, but not necessarily, generated by the masses as sources of the field, the way that charges are one possible source of the electromagnetic field (but not the only possibility, the other being a changing magnetic field). It is well known that general relativity does not actually satisfy the requirement that the inertial forces derive, somehow or other, from the totality of mass-energy in the universe. The general theory sets up an inertial structure in space-time which are the "natural force-free trajectories" or geodesics of freely falling bodies. These curved trajectories play the same role in relativity that inertial motions play in Newtonian physics. But they also define a kind of absolute structure in the sense that deviations from these trajectories, such as rotations and other non-uniformly accelerated motions, are detectable in the same way that deviations from inertial motion are detectable in Newtonian mechanics, for example in Newton's famous *Principia* bucket experiment and the experiment of the rotating globes (see Barbour and Pfister 1995). In fact, the natural trajectories deviate from each other in curved space-time, for example the geodesics of freely falling particles "cave in" toward each other at an accelerated rate as the particles fall toward the earth, and this is how the curvature of space-time is defined, by measuring this intrinsic deviation. Inertial structure is also present in the flat Minkowski space of special relativity even when there is no mass-energy present.

Einstein's theory of gravity, even if it was inspired by Mach's *Mechanics*, looks a lot more like a theory of absolute space-time curvature caused by mass and energy than it does like a Leibnizian–Machian theory of relative motion. Many philosophers use this argument to associate Einstein more closely with Newton and not with Leibniz or Mach. Another issue (raised by Kretchmann and by Friedman 1983, pp. 46–70) deals with Einstein's confusion over the mathematical meaning of general covariance. This purely mathematical property of the tensor field equations implies nothing about the physical relativity of motion, or physical invariances, of the theory, since even theories with absolute inertial structure, like Newtonian dynamics, have generally covariant formulations.

As a Mach-purist of sorts, I would simply point out (Banks 2003, 2012) that Einstein's redefinition of Mach's Principle, excluding the field and requiring as it does some kind of instantaneous gravitational interaction between very distant masses to account for the local effects of inertia, was very different from what Mach himself proposed, and the issue is hopelessly overshadowed today by Einstein's influence on the discussion. Even Julian Barbour's sympathetic reformulation (Barbour and Pfister 1995, pp. 218–220 and Barbour 2000) that "physics must use only relative quantities" is un-Machian in some ways. There might well be absolutes or physical invariants like fields in Mach's view, as above, so long as *motions* are relative. Mach's relationism permitted the idea of a *combined* inertio-gravitational field that is physically absolute, similar to the way we now think of the combined electromagnetic field as a real thing, invariant through the Lorentz transformations, except that the inertio-gravitational field would also be invariant through arbitrarily accelerated motions. The combined field, as Mach himself recognized, is certainly not an "absolute space" indicating a preferred state of rest or motion. The only difference between Mach's proposal of a combined inertio-gravitational field and Weyl's (1920, 1922), Einstein's 1953 statement (in the foreword to Jammer 1993), and Lanczos's (1970) idea of the gravitational field has to do with the acceptance of a field *without* source masses.

It appears from the 1883 quote that Mach was indeed prepared to accept an independently existing field and that the source masses were the only presently available reference frame for motions, not the only *possible* one. In 1920 at Bad Nauheim Einstein pointed out against Philipp Lenard that the presence of the source masses did not matter if we have field equations which describe the behavior of the field directly. This was also a point Faraday and Maxwell made when they originally introduced the field concept: namely that the field was a real physical entity in the space around bodies. As Einstein put it in 1953, in his so-called last testament in the Foreword to Max Jammer's *Concepts of Space* (1993, p. xvii), one has a choice between an absolute inertial system, which exists even in the absence of matter and energy, *or* one can accept the reality of a field without source masses but completely invariant to all states of motion as its gravitational and inertial components simply change places for observers who move relative to it.

To many scientists and philosophers of science, these distinctions between absolute fields and relative motions, and absolute motions or trajectories in space, make little difference (Rynasiewicz 1996) and maybe they are right. But the issue is crucial for Mach, since he also envisaged a radical "eliminationist" program for spatio-temporal concepts in physics, to uncover the underlying physical reality of events and their true causal–functional relations, of which his relationist program was only the beginning (Banks 2003, pp. 33–38; Norton 1995, p. 45). In setting this further goal, Mach is actually very close to Leibniz, who adopted an orthodox relationism in his debate with Newton and Clarke, but who used relationism about motion as a means to uncover the invariant physical reality prior to motion. Leibniz showed, in *Specimen Dynamicum* Part II (Leibniz 1989, pp. 130–137), that the force of percussion in the collision of two billiard balls (with the same mass) was physically invariant even in accelerated reference frames, whether we follow the accelerating ball or the decelerating ball, and thus there are indeed absolutes like the total overall force. But whether this force is viewed as "active" force, i.e., the kinetic energy of the striking ball, or as the "passive" forces of inertia and elasticity resisting acceleration in the ball that is struck, depends on the motion of the observer relative to the event. This surprising example may be the first use of inertial forces in physics, prior even to D'Alembert, who is usually credited with the idea. Curiously, Leibniz, like Mach, is often read as a pure relationist who denied absolutes of any kind in physics, whereas both men only seemed to deny absolute motions or states of rest.

Also like Mach, Leibniz imagined an explicit elimination of spatial and or temporal extension from physics in the late essay *Initia rerum mathematicarum metaphysica* (see De Risi 2007), which we may infer is his deeper view, not represented at all in his debate with Newton and Clarke. Mach also said many times throughout his physical writings that he believed that physical events and principles did not have to be interpreted spatially once they were reduced to a pure "functional" form. So whatever the outcome of absolute, or substantivalist, versus relationist debates in the philosophy of space and time, Mach's real goal seems to me to be the eliminationist program, and the relationist project, however famous and influential, is only a first stage.

I mentioned before that Mach thought physics needed to be set free from its historical path of development and that he also wanted to expose principles in physics like the law of inertia, the parallelogram of forces, and the law of the lever as empirical, not a priori, truths. This was one reason to strip the theories down to element and function form. But what was a "sense-physiological" critique of physics and why was Mach so willing to give up on the mechanical philosophy of nature, especially when Maxwell, Helmholtz, Boltzmann, Hertz, and Planck were bringing the science to perfection? In the usual telling of the story of nineteenth-century physics, the realists Planck and Boltzmann are the heroes, overcoming positivist skepticism about atoms, and further developing the kinetic theory of gases and the mechanical world view, Boltzmann finally "reducing" the second law of thermodynamics to ordered and disordered configurations of distinguishable molecules, and dispensing with the need for postulates of experience that were not backed up by a mechanistic explanation.

What Mach objected to in the mechanical philosophy, as well as atomism, however, was not its physics per se but its metaphysics. This means its reliance on mechanical concepts like mass, space, and time which Mach thought needed further investigation. "Metaphysics" was an idiosyncratic term for Mach, who did not take it from the history of philosophy but from his research in sense physiology and his reading of Kant. This is typical of the way Mach defined philosophical terms his own way. Mach strongly believed that the development of physical concepts, like mass, space, time, or force, was influenced by the human capacity to psychologically visualize characteristics of the material world. It is exactly this surprising ease of intuitive visualization of physical processes that made these concepts suspect for Mach. Why should nature be organized in such a way as to satisfy our psychological intuitions of objects? This critique of sense-physiological "metaphysics" in science (see Banks 2013a) sealed Mach's opposition to the mechanical philosophy and played a significant role in his critique of the basic physical notions to which he devoted his utmost scrutiny. Mach wrote in a 1910 retrospective article, to answer the criticisms of Max Planck, that despite their success, granted on all hands, these mechanical concepts are not the bedrock foundations of science but remained merely commonly understood unintelligibilities, and fundamental "problems" to be solved (Mach 1910). He added that he believed Einstein, Lorentz, and Minkowski were making progress on this front and said elsewhere that he subscribed to the theory of special relativity as developed by Minkowski in 1909 (Mach 1872/1910, p. 95). As we now know, he was getting tutorials in special relativity from Philip Frank (see Holton 1988).

Mach's thesis in the 1872 *Conservation of Energy* and the later 1883 *Mechanics* was that the most basic principles of mechanics and thermodynamics, such as the conservation of energy and the law of least action, were much broader than their mechanical application to systems of bodies in motion would suggest. Hermann von Helmholtz had championed the contrary view in 1847, i.e., that all natural processes would ultimately reduce to a mechanical description in terms of particulate matter acting on other matter by central forces producing motion. For Helmholtz, the reason the conservation of energy was true was because all systems reduced to classical mechanical systems at bottom. As Mach pointed out, however, the principle of excluded perpetual motion, one of the simplest ideas of all, was a key assumption required by mechanics, which could not be proved mechanically without making a circular argument. Helmholtz had simply assumed conservative forces in his description of mechanical systems (Banks 2003, pp. 184–185).[5] For Mach, the laws of mechanics were more like those of thermodynamics or electromagnetism—general "phenomenological" laws or postulates, resting on experience as their foundation and independent of the detailed mechanisms or mechanical systems in which they were realized (Mach 1883/1960, pp. 598–599, Banks 2003, pp. 164–169). The job of such general laws was to unify and derive more specific empirical laws under more abstract theoretical principles, to economically unify the specific under the more general patterns of natural events, not to lend credence to *any* particular metaphysical picture behind natural events (see the final section of Mach 1872/1910).

Mach's work illustrates perhaps a perennial divide between the empiricist and the realist philosopher of physics. The realist holds that for every law and regularity in nature there must be underlying entities, mechanisms, or causes that realize the law, or make it true. The motion of the hands of the clock is explained by the gears behind the clock face, not simply by cataloguing the movements in a list. The flow of heat and diffusion of gases are explained by the statistics of colliding particles, the second law of thermodynamics is explained as a probabilistic statement about the most likely arrangements of particles in a gas, and so on. The empiricist, on the other hand, readily admits all such cases when mechanisms can be verified, or established continuous with experience, but he does not make this the be-all-and-end-all goal of science. He need not enforce belief in any ultimate level or erector set of basic entities and mechanisms. Instead, it is possible for science to adopt general abstract theoretical principles or postulates of experience, which unify empirical laws of lesser generality, but which need not themselves be explained by mechanisms of any sort. Einstein's theory of relativity and the quantum theory are both postulate-based phenomenological sciences, for example, *not* mechanistic theories with realizing mechanisms.

These basic phenomenological laws stand on their own, governing natural events directly, and are not backed by any underlying theory of matter.[6] It is rather the reverse: the laws are used to predict properties

---

[5] Helmholtz moderated his views of physics later away from the simple mechanism that marked his earlier years.
[6] Some philosophers of science distinguish phenomenological or empirical laws from theoretical laws in a different way. For example, Snell's law of refraction is phenomenological not in the sense of being fundamental, but in the sense that it is a superficial law which can be derived from Fermat's

matter must have. Behind the seeming solidity and permanence of the visual mechanism is the greater solidity and permanence of a law or a postulate. For example, the empiricist understands the second law of thermodynamics as a general principle which *can* be realized in the disordered motion of molecules in a gas, but which outruns all of these realizations and is generally valid no matter what application may be made of it. Hence, for example, a Maxwell's demon introduced to contravene the second law by mechanical means (it is asserted) always ends up creating as much or more entropy as he reduces no matter *how* the demon is realized, showing that the law is more general than its mechanical realization. Carnot's deduction of the limit of ideal efficiency of heat engines is explicitly shown to be independent of the nature of the engine and what materials realize it.

Yet another illustration of this sort of phenomenological reasoning is Kirchhoff's deduction of a general natural law of black-body radiation. Kirchhoff showed that if two bodies are in thermal equilibrium, exchanging heat or radiation between them, absorbing and emitting it back and forth, a body's emitting capacity had to be equal to its absorbing capacity for energy, or else one body would keep some radiant heat energy at the expense of the other. But they are in thermal equilibrium, so one body would become warmer at the expense of one that gets cooler. This would violate the second law of thermodynamics since the net result would be a body gaining heat by the cooling of another body. The equality of a body's emitting and absorbing capacity *must* be true no matter what the constitution of matter and radiation may be, or what model may be used, and it set the stage for Planck's inquiries on the empirical law of black-body radiation, which could be considered a universal law of nature even when the structure of matter was still unknown. In fact Planck used the phenomenological black-body law for a theoretically *perfect* emitter and absorber of radiation, which was true for any kind of matter and radiation whatever, to deduce what the nature of matter and radiation was at the atomic level, not vice versa.

For Mach, then, establishing a thoroughly *phenomenological physics* is a first step in replacing a particular realization of the law with the more abstract framework of function and elementary event. Phenomenological laws of physics were candidates to become Mach's pure causal–functional

---

Law of Least Time for light. A phenomenological law in this sense would be an empirical regularity for which we have not discovered the underlying mechanism, entities, or the theoretical law from which it can be deduced. That is of course, not Mach's meaning.

connections when stripped down to fundamentals, stripped of mechanisms, visualizations, and even their spatio-temporal concepts, and when the empirical content of these laws was re-expressed in the form of concrete elementary events, observed and unobserved, to fill in partially observed experience and complete the functions even beyond immediate experience. Such was the absolute "bare bones" content of scientific laws and facts of experience for Mach. In this way, the purely phenomenological content of the law could be isolated from other theoretical assumptions, historical or a priori, and finally made independent of any realizing theory of matter or underlying mechanisms. Since these extras could well be wrong, he thought they should not be represented as part of the fundamental "breathing" content of the theory (Mach 1872/ 1910, pp. 59–74).

For Mach, all one can really know about, in, or even beyond present experience are the elements and their great variety of functional forms explaining their occurrence in events, laws which must be gleaned from experience. This is not a question of the depth of explanation: organizing experience into a catalogue simply because that is all one has direct access to. Behind appearances are just more appearances, needing to be organized in the same way. The reality behind appearances is one of more events and more functions, at present unknown but at least continuous with what *is* known as long as we are not fooled by visualizations and models or historically conditioned a priori prejudices. As is now obvious, what Mach was really offering his scientific colleagues was more than an epistemological organization of experience, or even a methodology of science. Rather, the element-and-function ontology is a true natural philosophy intended to compete directly head-on with the mechanical philosophy. It is also, clearly, a metaphysical view of its own, a nominalist, empiricist view of what is real, despite Mach's protestations to the contrary that he feigned no metaphysics.

Now, in the late nineteenth and early twentieth centuries, it did appear that realism and the atomic theory would triumph over empiricism by doing exactly what Mach had warned *against*, by realizing all laws of physics in the mechanical system of atoms in space and time, even the abstract postulates of thermodynamics, and readily embracing intuitive models and visualizations of natural processes. It was rather in the next two generations of physicists, including Einstein (born in 1879) and Heisenberg (born in 1901), where Mach's intellectual influence would be strongest, as the visualizations behind classical physics began to break down and one was left to contemplate abstract symbolic relationships one could not understand by "making a model" of what was going on. This of course was exactly what William Thomson once called a prerequisite to any scientific understanding of nature: could one make a visual, mechanical model of it? Surprisingly, we now know this is not the case.

Of course, one needs to be careful to restrict the Machian, empiricist influence to certain periods. While Heisenberg remained an empiricist, and a spokesman for the Copenhagen interpretation (with its own complications: see Camillieri 2009), Einstein began as a positivist and ended up a dyed-in-the-wool *realist* (Holton 1988, ch. 7) who opposed the quantum theory exactly because it did *not* give a visualizable description of separable systems in space and time, which Einstein apparently regarded as Kantian "preconditions" for the practice of science (Howard 2005). Heisenberg's matrix theory of the atom (see Heisenberg 1930) does, I believe, come very close to realizing the Machian umbrella schema of observable elementary events or transitions bound in abstract functions, or symbolic relationships, with an explicit denial of a spatio-temporal visualization for the underlying processes, like the coordinate of the electron and a path and even the very identity of the particle, if there is more than one. There is no concession toward visualization of a mechanical system in these laws, as long as they do not contradict observation and unify the event particulars that occur in experience and elsewhere in nature. The anti-empiricist tendency still current in the history and philosophy of science today is to downplay Einstein's and Heisenberg's early empiricism because of these later developments in their views. It is true that certainly after their initial breakthroughs, when both men worked with schematic and pared-down empirical theories, they moved subsequently in a more realist direction, Einstein perhaps in the direction of Kantian "empirical realism" about space and time, and Heisenberg in the direction of "Aristotelian" realism about *potentia* (*dynamis*) (see his 1986). It should not, however, be denied that their empiricist "fresh view of the facts" (facts which other scientists also knew but failed to see in the proper way) was of first importance in establishing the entering wedge for the later theory *and* its realistic interpretation.

Conclusion

Clearly, we are long, long overdue for a reconsideration of Mach's philosophy and science. One of the most overlooked features of Mach's work was his realistic empiricist theory of elements and his overall scheme for empiricist theories, none of which had any relation to sensationalism or other forms of idealism, or any deep conceptual links to the Vienna Circle and its style of empiricism. If only these facts about Mach were more widely known, I believe the old stereotypes would quickly crumble and we would be ready to reintegrate Mach into discussions in the philosophy and history of science where he always belonged.

CHAPTER 2

Mach's philosophy of mind

Introduction: neutral monism in psychology

Mach was also a major figure in the psychology of sensation and perception, but he kept these interests separate from his physics for the most part, designating physics and psychology different departments of inquiry, for the present, about different kinds of functional connections among the elements (1872/1910). As in the case of his reform program for physics, I think of Mach as proposing a *schema* for a family of realistic empiricist theories of perception. Again, his point was to say what such a theory should look like, once psychology had been freed from a whole raft of metaphysical "pseudoproblems" preventing it from reaching the status of a science. This reform program runs more or less parallel to Mach's critique of physics. Once again, the goal is a stripped-down theory schema that can be used as a pattern for future theories.

Mach developed his ideas in psychology in the 1860s, deeply influenced by the philosophy of J. F. Herbart and especially by the appearance of G. T. Fechner's *Elemente der Psychophysik*, even if he had given up Fechner's famous psychophysical law by 1865 for his own theory of neural inhibition (see Heidelberger 2004, Banks 2003, ch. 4). What was good about psychology for Mach was the direct concrete evidence of existence it provided of the natural world through sensation. This possibility of some kind of direct, concrete contact with reality, however complicated, explains the appeal of psychology to empiricist philosophers and it is the one area where psychology is actually superior to physics in giving us what Mach called a "real knowledge" of nature (1905/1976, p. 361). Yet without a theory design, such knowledge is confused, jumbled, and clouded by metaphysical errors, in comparison with the precision of physics.

Mach was well aware that conscious mental events are a highly restricted set of events in the brain, most of which are non-conscious. The great majority of events in the brain are like the bulk of the iceberg under the surface, of cells and their configured structures, which support the small minority of conscious events and functional connections which go on inside these structures, and are only artificially separated from them as a group. A psychological theory which confines itself to functionally relating *only* the elements present to consciousness is bound to be superficial. As Mach was well aware, for example in his work on the Mach Bands and other sensory phenomena exhibiting non-conscious processing (see Mach 1886/1959, p. 71, Ratliff 1965, Banks 2001, and especially Pojman 2000), the actual conscious sensation, and its functional relations to other conscious data, is only the final layer of the many processes that precede it, structure it and, as Mach says, "hand over the final result to consciousness" (1886/1959, p. 71). Thus, the conscious mental events to which we have access were certainly real for Mach, but they were also a kind of superficial top layer prepared for by these many complex physiological processes in the brain which precede them. A psychological theory of the elements can certainly identify functional connections between conscious sensations, images, and cognitive processes among themselves, but every psychological theory, for Mach, is also anchored in that iceberg of non-conscious physiological brain processes. Hence Mach's staunch allegiance to a "physiological psychology" tied into events in the brain, *not* a purely phenomenal or introspective psychology dealing only with conscious elements (1886/1959, pp. 60–71).

Elements admit of a wide variety of natural variations, which investigators may choose to specialize into the mental and the physical, but which do not represent hard and fast boundaries, either between the contents or the functions that relate them. Mental "functions," or associations, linking a sensation to mental images or feelings, simply reflect the underlying physiology of the brain or the sense organs for Mach (1886/1959, p. 62) and are therefore something physical, in an enhanced sense of "physical," but there are no truly separate psychological laws. Even mental images or hallucinations are simply facts about internal energy changes in the nervous system and ultimately are purely physiological in nature, neither true nor false of anything. They simply are (1886/1959, pp. 10–11, 199–202). There is nothing about either the content or the functions that is irreducibly "mental" in any way. Any sensation (s) can thus be labeled just as easily as a physical element (e), or as neutral between the two (s/e): see Figure 0.2.

In the *Analysis of Sensations*, Mach tries his utmost to restrict his discussion to these observable 's/e's. He readily admits that the sensations of other people and animals must be added to one's own, as well as the unobserved elements filling out objects, such as the backs and sides of chairs and buildings in three-dimensional space, or the missing past and future positions of a body moving in a parabola. Mach also says that we can infer the causal behavior of unobserved elements in objects, continuous with observation, by comparing the object's appearance for a sequence of difference observers. If, for example, A, B, C . . . are sensation/elements of a penny to one person S, and A′, B′, C′ and A″, B″, C″ are sensation/elements of the penny to two other observers, S′ and S″, it will be possible, by eliminating all of the variations due to the peculiarities of S, S′, S″ . . . and so on, to isolate what Mach calls certain "spatial identities" of the penny (its position, its size, and perspective, its relation to other objects) set free from each observer individually and invariant across a variety of perspectives (Mach 1886/1959, pp. 344–345). William James (1977, pp. 209–211) proposes a similar criterion for perceptual objecthood as the sum of all the subjective perspectival views of the object. If a group of observers all point at an object, each of the others will seem to be pointing in a different direction, from his own egocentric point of view: what is dead ahead for him is at an angle for me. But, says James, the location of the object is still to be found in an objective space where all our hands meet.

But this is clearly a sort of unsatisfactory halfway-house position, proposing an "observer-dependent/observer-independent" object which remains constant across all acts of observation, but must still be observed in order to exist. Ultimately this proposal does not serve to ground the mind-independence of even the object's primary qualities like its location or extension, as Berkeley argued quite convincingly. All we know is that the objects have some properties that remain the same across all subjective perspectives, a kind of subject-dependent/subject-independent set of properties. But we do not have any evidence that these properties would exist if all of the observers S, S′, S″ . . . were absent.

Mach of course *did* believe in fully mind-independent elements and complexes, as did Russell. Sometimes Mach said he was only interested in the kind of elements that can also be considered sensations under a different ordering, the **s/e**'s. Other times he said that he recognized elements that were never anyone's sensations under any interpretation, in the case of other minds, animals, and events in inanimate matter:

> To that which I observe, to my sensations, I have to supply mentally the sensations of the animal, which are not to be found in the field of my own sensations. The antithesis appears even more abrupt to the scientific inquirer who is investigating a nervous process by the aid of colorless

abstract concepts, and is required, for example, to add mentally to that process the sensation green. This last may actually appear as something entirely new and strange, and we ask ourselves how it is that this miraculous thing is produced from chemical processes, electrical currents and the like. Psychological analysis has taught us that this surprise is unjustifiable, since the physicist is always operating with sensations. The same analysis also shows us that the process of mentally supplementing complexes of sensations according to analogy by means of elements which at the moment are not being observed, or by elements which cannot possibly be observed, is one which is daily practiced by the physicist; as, for example, when he imagines the moon a tangible, inert, heavy mass. The totally strange character of the intellectual situation above described is therefore an illusion. (Mach 1886/1959, p. 43)

Notice that Mach is saying that we may, if we wish, start with the whole array of elements and work our way in toward the individual ego, as well as starting from the sensations present to some particular ego like our own and working our way out, completing missing elements and continuous functional relations until we find ourselves in the general array again. He sees no difference between these two cases except a different focus of inquiry, not an insuperable metaphysical gulf between the content of a particular mind and the rest of nature and other minds since ultimately all the elements exist together in one fabric (Mach 1905/ 1976, p. 361).

In my previous book (Banks 2003, Introduction, chs. 7 and 9) I have cited direct evidence from Mach's writings in which the world elements are mentioned explicitly. He says, for example, in a notebook fragment from 1872 that "sensation is a general property of matter," and even in later writings he spoke about a stage where the world elements in matter could be accessible to experiment and direct verification, "when the tunnel is built through," referring to the day when mind-independent elements could be directly experienced. These passages match quotes from letters to Friedrich Adler and Gabriele Rabel, who in turn cited them as evidence that Mach believed in realistic elements in matter. Mach himself, in *Knowledge and Error* (1905), even described direct knowledge, or experience, of these world elements as "real knowledge" of nature, not the mere "spatial identities" mentioned above:

If the ego is not a monad isolated from the world but a part of it, in the midst of the cosmic stream from which it has emerged and into which it is ready to dissolve back again, then we shall no longer be inclined to regard the world as an unknowable something and we are then close enough to

> ourselves and in sufficient affinity to other parts of the world to hope for
> real knowledge. (1905/1976, p. 361)

This matches a similar passage in the *Mechanics*:

> Careful physical research will lead to an analysis of our sensations. We shall
> then discover that hunger is not so different from the tendency of sulfuric
> acid for zinc, and our will not so different from the pressure of a stone as it
> now appears. We shall again feel ourselves nearer nature without its being
> necessary that we should resolve ourselves into a nebulous and mystical
> mass of molecules, or make nature a haunt of hobgoblins. (1883/1960,
> p. 559)

Bold ideas, to say the least. John Blackmore (2006) has recently taken me
to task for citing such utterances of Mach from his notebooks proving that
he accepted mind-independent world elements, some of which were
already published by Haller and Stadler (1988). Blackmore sees these
passages as mere after-thoughts and additions to a position that is overwhelmingly phenomenalistic and positivistic, and he believes they represent mere stages and not consistent features of Mach's mature position.
I disagree on both points. First, there is enough evidence from Mach's
published remarks to infer the existence of world elements; the unpublished
remarks simply confirm what is already there. Second, Mach held that
even directly experienced sensations are already as physical a part of the
natural world as anything can be, and thus he saw no further intellectual
hurdle in assuming that these (s/e) elements stood in continuous causal
relations to other world elements (e) in nature. We should therefore always
begin from the complete array of elements making up the world, as in the
last two quotes, and restrict them only for the provisional needs of special
inquiry. Here is another passage from the *Analysis of Sensations* where
Mach takes this "overall" view:

> Now if we resolve the whole material world into elements, which at the
> same time are also elements of the psychical world, and, as such, are
> commonly called sensations; if, further, we regard it as the sole task of
> science to inquire into the connection and combination of these elements,
> which are of the same nature in all departments, and into their mutual
> dependence on one another, we may then reasonably expect to build a
> unified monistic structure upon this conception, and thus to get rid of the
> distressing confusions of dualism. (1886/1959, p. 312)

Notice that Mach is speaking here of a general array of elements, which is
only divided provisionally into mental sensations and elements making up
physical objects by concentrating on a different class of natural variations.

In *Knowledge and Error* (Mach 1905/1976, p. 361) Mach calls attention to the mistake of establishing hard dividing lines inside one continuous fabric of experience-reality, calling some of it the ego and the rest of it the world, where no such boundary is indicated, or coherently conceived, since drawing such a boundary means already standing outside of it.

Mach believed this illusory ego–world boundary had been crossed already when he posited that a sensation like a red patch was *already* part of the physical world. If these sensory qualities, like red, are already concrete physical, or natural, events in another set of variations, then we are already outside the bounds of the ego and the sensations can then be causally related to other physical events in mind-independent nature. The problem is not to "break out" of the ego toward the external world, à la Descartes, like trying to reason our way out of a prison cell. We are *already* outside that boundary when we consider sensations to be physical events that are part of the mind-independent physical world under a different set of variations. So actually realism about the external world is established already, without Mach's needing to invoke completely mind-independent world elements; yet he clearly believed in them too, to "fill out" the rest of the overall array that are not anyone's sensations under any set of variations they undergo, the pure **e**'s in Figure 0.2.

Sometimes Mach begins an investigation with the whole array of elements, in which egos are embedded like little islands (in the *Conservation of Energy* or the *Mechanics*), and sometimes he adopts a point of view within one ego and shows that one's sensations are already physical events and hence non-mental (as in the *Analysis of Sensations* or *Knowledge and Error*) so that we are already outside the ego in the mind-external world. Mach saw these viewpoints as entirely compatible, as indeed they are.

The functional composite ego

Mach's famous deconstruction of the ego ran along similar lines, which would be picked up later by William James in his "Does 'Consciousness' Exist?" Where other philosophers and psychologists like Franz Brentano saw an essential unity of consciousness underlying all acts and experiences of the mind, and its contents, Mach saw only the unity of a "function," bringing together many other automated sub-functions, as it were, many of which Mach had explored in his work in psychophysics in the different senses. Mach viewed the mind as a composition of many separate mechanisms, each with a different task and each, like some kind of little

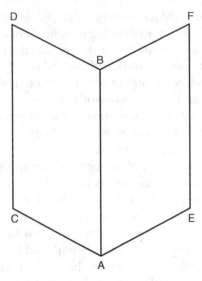

Fig 2.1   The Mach visiting card

machine, passing on its results to some other functional mechanism for further processing. "Consciousness" for Mach is just the sum total of all of these activities composed together as functions are composed inside of one another. In fact, Mach focused his attention on processes in perception that seemed almost automatic and not under conscious or unconscious control of any centralized ego: Mach Bands for contrast phenomena, depth, light perception, and mechanisms of color composition (see Banks 2001). One simple example is Mach's explanation of his "visiting card" illusion (see Figure 2.1).

A card is bent in the middle like a roof and placed under strong light so a shadow appears on the right side. Then we voluntarily reverse our depth perspective, seeing the card folded like a book. Immediately, the shadowed side of the card appears to be painted grey and bathed in strong light and the line AB seems tilted away at an angle as depth and orientation change. I have even reported a third possibility (Banks 2001) where AB appears at a uniform depth, but the panels contract into flattened trapezoidal shapes. Hermann von Helmholtz, a great authority in matters of perceptual psychology, would no doubt have explained this illusion as an "unconscious inference" by a central ego, to the effect of "that looks like a roof of a house" or "that looks like an open book before me." But to Mach the depth reversal is rather cued to the sensations of light, length, and orientation, more or less automatically, like the parts of a machine, and no higher-order judgments are necessary. Indeed, for Mach, one is programmed to see the card in these various ways because of innate evolutionary mechanisms and not inferences from past perceptual experience of objects like roofs and open books. These innate mechanisms underlie our judgments about objects. What is interesting about this illusion is that Mach sees all functional connections in play, between sensations of light, depth, form, and orientation. Any type of sensation can enter into a causal–functional relation with any other type, depth with light, orientation with form, what have you, relations at which the conscious mind might balk if the types are too dissimilar.

Also, Mach did not believe sensations got their concrete quality or existence from being apprehended by the mind in a fundamental "act" of conscious representation. He recognized that all sensation/elements occurred in complexes of one sort or another, and one kind of complex could be described as a circumambient awareness of particular experiences. For example, surrounding events with other events in functional relations, memory images annexed to them in time and space, could account for some of the features of awareness or attention. But *all* natural elements occur in complexes of some sort and this does not mark off anything especially "mental" about them. The elements and their qualities as individual events do not depend on the functions they are in, it is rather the opposite: the function depends upon, and is grounded in, the qualities of the events it relates (Mach 1905/1976, pp. 31, 32).

This Machian view of consciousness stands in direct contrast to the views of some philosophers, who hold that consciousness is the mark of "qualia" or having experiences such that there is "something it is like" to have them, as if it were the circumambient consciousness, act of representation, or awareness, that gave individual experiences their concrete quality and existence (see Harman 1990, Strawson 2006). Mach believed the various sub-functions composing what we call conscious activity—linking sensations to memories in a time order, emotions, mental images, spatial orderings—were indeed real and ultimately linked solidly with physiological functions of the nervous system. All psychology eventually would be reduced to physiological psychology, the more the nervous system was understood. The impulse to believe that qualia represent something non-physical arising from representational acts or abilities unique to minds, or that the quality of a sensation/element depends on its being apprehended by a consciousness, was totally rejected by Mach, *avant la lettre*. As he succinctly put it: "Das Ego ist unrettbar," the ego cannot be rescued. What remain, as in the case of the body, are the elements/sensations and functions. There is no need for an "owner" to precede these contents:

> If a knowledge of the connexion of the elements (sensations) does not suffice us and we ask, *Who* possesses this connexion of sensations, *Who* experiences it? Then we have succumbed to the old habit of subsuming every element (every sensation) under some unanalyzed complex, and we are falling back imperceptibly upon an older, lower, and more limited point of view. (Mach 1886/1959, p. 26)

Mach's physiological view seems to me to be a lot more in line with actual neuroscience and the surprising findings that such intimately related processes as shape and motion perception can be separated, phenomeno-logically, as well as physiologically, from the unity of consciousness, and even combined arbitrarily in some cases to fuse into different kinds of perception. Seemingly "impossible" perceptions such as perceiving motion without a thing that moves, or an object without a definite color or shape, *are* actually possible and explainable on the idea that there are separate functions for color, shape, and motion composed together in perception. Perhaps even more stunningly, there is no physiological evidence for a unified center of consciousness at all. In the case of the visual field, for example, there are layers and layers of processing of the retinal image to pick up a variety of features, but there is no viewer looking at his own visual field as if on an internal screen, there is only the visual field itself mapped onto the visual cortex. Our introspective evidence for a unity of consciousness can be thus drastically wrong and misleading, just like our language for talking about mental processes introspectively.

Mach was well aware of the competing view in Austrian philosophy, stemming from Brentano (see Smith 1994). According to Brentano, the mark of the mental is "intentional inexistence." An act of the mind, such as perception or memory, is intentional, Brentano says, if it somehow con-tains the object or is about the object, even if the object is not physically present. Thoughts and other mental acts thus have the ability to refer beyond themselves to transcendent contents and objects, a property of "aboutness," that nothing in the physical world has. The evidence for this thesis is, again, introspection; we supposedly sense this intentional prop-erty of mental acts by an act that is *likewise* intentional. In this way, Brentano and his followers claimed, we could sense not only the red sensation but also the *act of the seeing* of red. We do this via yet another introspective act of seeing "the seeing of the red," and so on. James and

Russell would later directly challenge Brentano's "evidence" and find that in fact, no such internal perception of the act *was* verifiable. James's colleague C. S. Peirce was even more skeptical, famously pointing out that the only evidence for the "power" of introspection was based upon introspection itself (Peirce 1868). Verificationist arguments like this seem to have convinced James and Mach to give up on the fundamental intentional act theory of Brentano and to replace it with the view that consciousness is the functional composition of many other simpler sub-functions. As we shall see below, the case of Russell is a bit more complicated. He too cited this verificationist argument of Mach and James but it seems not to have been the decisive turning point for him in abandoning his theory of acquaintance and embracing neutral monism.

"Sense data" theory, "Myth of the Given"?

Many philosophers or psychologists who would grant Mach's view of the ego would still fault him for holding an atomistic "sense data" theory of sensations. This is another easy charge to refute. Mach, in the *Analysis of Sensations* and an earlier 1871 essay called "On Symmetry," explicitly called attention to what he called sensations of overall form. We not only have sensations of the individual points of a circle, we also have overall sensations of its roundness and its rotational, reflective, and other symmetries, due, Mach thought, to the symmetry of the eye muscles and visual system through which the contents are viewed. These "higher-order sensations" are thus linked polyadically to clusters of the individualized sensations of points, the way for example that the form of triangularity can be sensed of individual dots or edges arranged in a suggestive pattern. Christian von Ehrenfels, one of the founders of Gestalt psychology, credited Mach with originating this idea of form sensation, one of the founding notions of Gestalt psychology, hardly a school of psychological atomism.

Some might also see Mach's sensations as illustrating Sellars' Myth of the Given (Sellars 1956/1997). Sellars sets up a dichotomy by saying first that "givens" in experience must be simple if they are to be epistemically fundamental. But if they are simple, nothing can be derived from them as they have no internal structure. If the "givens" are complex, we can derive other facts from them epistemically, but then they cannot be considered "given" in experience, since they are not simple. I suppose Sellars was thinking of empiricists like Mach when he framed his argument, but Mach's elements were not "atoms" or little segmented entities; they always occurred in complexes from which they must be empirically isolated.

Nor were Mach's elements ever structureless or lacking in internal complexity. Mach emphasized that seemingly atomic sensations can always be analyzed into simpler sensations, as colors can be analyzed into the set of primaries and as complex sensations of sound, position, and movement can often be analyzed into simpler components. This process never ends. Thus the sensations themselves are only "provisional," as he says, and their individualized boundaries and limits change with our ability to analyze them further. They are not the "givens," of inquiry, they are the subject of further scientific analysis, which begins with the complexes and not the elements (Mach 1905/1976, p. 12n). The direction of progress is thus *not* to build a logical structure from atoms, or givens, but the analysis of complex sensations into simpler ones, the "analysis *of* sensations" (as in the title of Mach's book). *Never* did Mach seek the kind of construction Carnap sought in the *Aufbau* or any kind of foundationalist epistemology utterly out of line with Mach's own fallibilist, naturalistic view of scientific knowledge.

Why, then, is Mach associated with "atomistic sense data" and or the "Myth of the Given" in the first place? Paul Feyerabend pointed out that Mach had been misunderstood by the logical positivists who claimed him as a forerunner for their own obsessions with protocol sentences, logical sense data constructions, and conventional empiricist thinking foreign to Mach's views (although it needs saying that Feyerabend himself may have misunderstood the logical positivist project: see Friedman 1999, Richardson and Uebel 2007, Uebel 2007):

> The first and most noticeable change is the transition from a critical philosophy to sense data dogmatism. Elements are replaced by sensations not just temporarily and as a matter of hypothesis, but once and for all ... Secondly criticism of science is replaced by logical reconstruction, which in plain English is nothing but a highly sophisticated form of conformism. As a result the idea of logical reconstruction became conformist. The task was now to correctly present rather than to change science. (Feyerabend 1970, pp. 179–181)

Feyerabend is completely right on one point, that, unlike the project of logical positivism, which was directed at the second-order language and structure of science, Mach's theory of elements was a first-order project, designed to constructively criticize and contribute to science by means of actual investigations of the real elements of nature, framing theories in both psychology and physics for further investigation. The logical positivists just had no interest in the ontology of the world, leaving this entirely to physics, and looking to philosophy solely to settle linguistic questions of how to *talk* about the world and how to describe the structure of scientific theories and explanations. *They* were the ones who saw the data as atomic facts, o-sentences, and such, relative to their abstract theories and analytic linguistic frameworks, where their true interests lay. The atomic data or givens provide the logical positivist with the empirical content suitable for a verificationist "basis" for theories of the universe but beyond that, he has no interest in them. Mach never saw the elements this way. For Mach, elements *are* the real constituents of the universe which we are always trying to study further.

Theory structure: causal–functional maps

With any dualism of mental sensations and physical elements disposed of, Mach was ready to order all of the elements, element/sensations, and functions side by side in one general array, which he called a causal–functional presentation, and which I will call a causal–functional map. A causal–functional map is a kind of a graph in which elements of any kind may be entered and laid down alongside one another according to their particular functional relations, in as many dimensions as is necessary to capture the multiplicity of these relationships. Every individual happening occupies its own node in the structure, different from all of the other individual happenings. The causal map is abstract and need not represent objects in either physical or psychological spaces. Rather, both kinds of space can be recovered from the map by selecting certain elements and relations from it. In particular, Mach said, there was no problem laying down the sensations of different egos in the same display with elements of the brain and other physical events:

> When I speak of the sensations of another person, those sensations are, of course, not exhibited in my optical or physical space; they are mentally added and I conceive them causally, not spatially, attached to the brain observed or rather functionally presented. (Mach 1886/1959, p. 27)

> From the standpoint which I here take up for purposes of general orienta-tion, I no more draw an essential distinction between my sensations and the sensations of another person than I regard red or green as belonging to an individual body. The same elements are connected at different points of attachment, namely the egos. But those points of attachment are not anything constant. They arise, they perish, and are incessantly being modified. (1886/1959, p. 361)

Nor was there any difficulty adding the purely physical elements to the array, for these are causally homogeneous with sensations and can enter into causal–functional relations with them directly. This, finally, is that one "monistic" fabric of experience-reality Mach spoke of above—with one sort of element and one sort of causal–functional variation—which he said he always took up for purposes of orientation in science, which did not need to be changed when passing from psychology to physics and back again.

The variety of causal–functional relations between the elements is very broad, from simple one-to-one reciprocal associations, to polyadic relations, annexing one element causal–functionally to a group of others, to relations of mutual disappearance, in which the occurrence of one element precludes the occurrence of another, so that when one appears the other promptly disappears. In the *Conservation of Energy* (1872/1910, pp. 70–71), Mach even discusses what he calls causal "islets," where large groups of elements abutting on one another form practically closed systems, such as objects or egos, or even sub-systems within them. There could even be higher-order functional relations among these systems of elements themselves, for example relations of "macro-causation" from system to system (see for example Mach 1886/1959, p. 55n, and Banks 2010, pp. 177–179).

The causal–functional graph is thus quite ecumenical but the variety of functional variations seems almost bewildering. The problem with his view, which Mach does not address anywhere, is exactly this over-loose flexibility of the function concept. What makes the function concept so useful and flexible is also its Achilles heel. We seem to have too much freedom and not enough stringency in our choice of functions. Although Mach himself does not make this point as explicitly as one would like, I will say it for him that his functions, as I understand them, are grounded in the manifested qualities of the elements themselves and their causal relations *qua* forces, so perhaps the variety of functions is really not so broad after all. We could perhaps read from the qualities and their changes what functions are candidates for unifying them under laws, and if there is more than one set of functions grounded in a given set of qualities, then so be it, but since the functions have to track variations in natural quality that are not arbitrary, there may be a natural limit built in for the variety of functional forms to find within experience-reality. The method is thus flexible enough to allow for more than one system and yet stringent enough to avoid the "anything goes" objection.

As it turns out, I think, and as I will explore further below, sensations and perhaps other mental events should be considered a species of these higher-order "macro-causal" functional relations and linked to islets of activity, and the qualities of our sensations should probably be linked to those macro-causal systems. Mental variations should be grounded specifically in macro-causal powers and relations. That is only a sketch, of course, which will have to be filled in later in Chapter 5, but it seems to follow from Mach's (1872/1910, pp. 70–71) suggestion that we can consider the formation of higher-order macro-causal "islets" among the elements (for example the ego) and Mach's idea that the ego may therefore be a "determining cause" of events just like other physical causes, consistent with the conservation of energy (1886/1959, p. 55n).

Mind–body relations in causal–functional maps

For Mach, causal–functional maps are the appropriate framework for mapping out psychophysical relations, and for framing reductive, physiologically based, psychological theories, which are otherwise fraught with what he calls the "metaphysical" errors and "pseudoproblems" (*Scheinprobleme*) of psychology, similar to those he had identified in physics. The most important relation to be analyzed, of course, is that between sensations and brain tissue, or certain complex events therein, with which sensations are identical. The subtitle of Mach's *Analysis* promises an explanation of "the relation between the physical and the psychical" as one of the main claims of the book. Of course, it seems that these are utterly different kinds of entities. When a person looks at a green object, like a leaf, his brain does not look "colored," to us, nor does it seem that qualitative sensations could possibly reside inside brain tissue, or an individual neuron, its constituent atoms, or even within a cluster of electrochemically signaling neurons. Either one sees this as some kind of insuperable problem for neuroscience, or one can see it the way Mach does, as a mistake or confusion in the statement of the problem, as follows:

> When I see a green leaf (an event which is conditioned by certain brain processes) the leaf is of course different in its form and color from the forms, colors, etc., which I discover in investigating a brain, although all forms, colors, etc., are of like nature in themselves, being in themselves neither psychical nor physical. The leaf which I see, considered as dependent on the brain-process, is something psychical, while this brain-process itself represents, in the connexion of its elements, something physical. (Mach 1886/1959, pp. 61–62)

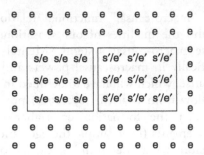

Fig 2.2   Three classes of sensation-elements

Mach believed this puzzle could be solved directly by carefully separating out the different elements in play, and entering them into the causal–functional map in their proper places. The element of green will be found to be causally related to other separate elements discovered by externally measuring events in the brain, and it will also be related to the elements of the leaf. When we make our causal map of the elements involved, we have to distinguish three classes (see Figure 2.2). First are the sensation/ elements **s/e** of a person, his actual concrete sensation of green, for example, which is neither mental nor physical. The second class (possibly overlapping with the first) consists of the physical elements of brain tissue in themselves **e**. The third class consists of elements causally annexed to the brain tissue either by impersonal measuring devices, also **e**, or other human observers with their own sensation/elements **s'/e'**. Each element, as an individual occurrence, is separated from the others and the mesh of causal–functional linkages are individualized, connecting each occurrence to each.

Let's begin with one anchor point: the sensations themselves, such as green. In their physical interpretation, they are the internal energies of the brain *an sich*, *not* the elements of the observed brain measured by the scanner, or represented to observers looking at the brain from outside. The sensations of green or blue **s/e** are thus identical with certain elements of the brain tissue, as they are in themselves, not observed from the outside, perhaps energetic events occurring in a complex tangle of neurons. Even if we have a confused apprehension of these brain events, it remains true that when we look at our sensations we are indeed looking at events in the interior of our own brains, and therefore at something physical. This is no longer true when someone *else* looks at our brain tissue or measures its physical properties with instruments or other interactions. These interactions produce other elements which go elsewhere on the graph. We can try to observe our own brains in the mirror, while we are seeing the green leaf in front of our eyes, but we will still never succeed in looking directly at the actual brain tissue where the sensation of green is currently taking place, since we are already looking at it directly in front of us on the leaf and the same individual event cannot be in two places at once on the graph. If I do somehow succeed in placing an electrode on just the right spot in my brain where the green is presently occurring, and produce a physical element, the sensation immediately disappears. All of these relationships, including the mutual disappearance, or replacement of one event by another, are causal–functional connections which will be marked out clearly on the graph, presenting no further problems.

To show the power of Mach's simple technique, it might be interesting to push a little further and experiment with the idea. Suppose, for example, that sensations **s/e** like green are events that arise in clusters of 1,000 neurons of a specific anatomical sort, wired up in a certain kind of functional pattern of firing (which seems plausible enough; see Kandel *et al.* 2000). We can imagine conducting an experiment in which we interfere directly with these 1,000 neurons, say by sticking 1,000 microelectrodes into them and siphoning off their neural energy as they fire. In this case, the sensation of green vanishes and is immediately replaced by elements $e_1 \ldots e_{1,000}$. When we take out the electrodes the sensation of green reappears. Hence we can say that the sensation/element **s/e** is linked by a polyadic function to the whole set of elements $e_1 \ldots e_{1,000}$, and the nature of the function is this, that when one appears the other vanishes. And since these elements cannot co-occur they go on different places in the causal–functional map. This doesn't mean the sensation/elements **s/e** are non-physical, or non-identical to neural events. On the contrary, mutual disappearance actually proves identity in this case, since it means the same energies that manifest in the sensation of green **s/e** manifest in $e_1 \ldots e_{1,000}$ under different experimental circumstances.

What the neuroscientist in fact measures in a scanner, while the subject is simultaneously sensing green, is some external effect linked indirectly to, but not precisely the same as, the neural energies that are presently manifesting to the patient in his own brain. In microelectrode recordings the electrode is always near the cell recording secondary effects, not siphoning off the neural energy directly. A doctor might point to a scan to explain why a patient is seeing spots or flashes in his visual field. He doesn't literally mean the spot in the visual field *is* the object pictured in the scan, but rather that the scan is a manifestation of individualized effects caused by the same cells that also produced the spot as a collective manifestation of their electrochemical activities, but which here are observed interacting with the instruments, **e**, or with the observer (**s′/e′**). They appear different because these manifestations *are* different. But the powers possessed by those cells in their electrochemical signaling and their arrangement and anatomy and so forth are still *identical* in both cases, which accounts for our equally strong intuition that we are, in fact, dealing with the same thing. The electrochemical neural energies in brain tissue are the same whether they manifest as sensations or as events registered on the 1,000 electrodes, but the manifestations are different for different events, sensing directly and externally interfering or scanning.

That is the whole substance of the sensation problem on the causal map treatment, about which still more will be said in Chapter 5. But the essential features are already present in Mach's *Analysis of Sensations* more than one hundred years ago. In particular, the experimental vanishing of the sensation when the neural energy is channeled into other interactions is the best available proof that sensations *are* physical, not mental, if that word is supposed to mean "not-physical," or separated from the physical world. It seems to me Mach's causal–functional presentation solves the puzzle nicely by putting different individual manifestation events in different places on the graph, but through their mutual disappearance it affirms the underlying identity of the powers that manifest in different and seemingly incompatible ways.

Metaphysical "pseudoproblems"

There are two related metaphysical "pseudoproblems" Mach himself thought causal maps would be especially useful in solving. We can call these (1) "the error of introjection" and (2) "the error of bilocation." Both depend upon a confusion between psychological space, made up of space sensations and sensations of bodily surfaces such as the skin and the retina **s/e**, and a physical space of physical elements making up objects in space and time **e**. The causal map itself, which includes bodies and minds within it, is an array of events that is not necessarily a *spatial* representation, as Mach emphasizes, by saying that the elements are not spatially but "functionally presented" (1886/1959, p. 27). The spatial representations of sensory manifolds on the one hand, and physical objects on the other, are both constructed from the more abstract elements and functions on the causal map. This multidimensional map is also rich enough in multiplicity to accommodate *two* such spatial constructions, one for psychology, which is composed only of **s/e** elements, and one for physics, which consists of the same **s/e** elements, now under their interpretation as **e**'s, and the pure **e** elements, which are not **s**'s under any interpretation.

The "error of introjection" was originally pointed out by Mach's colleague Richard Avenarius, who deserves equal mention as a founder of realistic empiricism. Avenarius pointed out that when we investigate a brain as external observers, we never find little colored sensations inside the brain tissue. But we also believe that sensations *are* somehow spatially inside the brain we are investigating, so we "introject" a little colored image spatially onto, or into, the brain tissue somewhere or other, and this is what creates the problem. Under this preconception, we are tempted to say ridiculous things, like "this observed brain tissue is colored," in defiance of all observational facts. Or one might say instead that we can't see them but "the atoms of the brain tissue themselves contain little proto-colors hidden inside them." Avenarius insists that these conclusions rest on a mistake, due to a faulty way of conceiving the problem. The brain tissue observations are a set of causal interactions with the actual brain that show up on our measuring devices or that are apparent to observers looking at the brain (the **s'/e'** elements). The subject can even observe his own brain in a mirror if he wishes. However this is organized, the sensation of, say, green will appear, to external observers, and even to the self-observing subject, *alongside* his own brain tissue, not spatially inside of it. He can even look at the green **s/e** and at his own observed brain tissue **s'/e'** at the same time; thus, these appearances never physically coincide. It is thus quite wrong for observers to "introject" sensations into the same brain they observe externally, even when they are both the external observer *and* the subject of observation. Mach pointed this out also in the rest of the quote given above:

> I once heard seriously discussed "How the perception of a large tree could find room in the little head of a man?" Now, although this "problem" is no problem, yet it renders us vividly sensible of the absurdity that can be committed by thinking sensations spatially into the brain. When I speak of the sensations of another person, those sensations are, of course, not exhibited in my optical or physical space; they are mentally added, and I conceive them causally, not spatially, attached to the brain observed, or, rather, functionally presented. When I speak of my own sensations, these sensations do not exist spatially in my head, but rather my "head" shares with them the same spatial field (Mach 1886/1959, p. 27).

In other words, the "head" (in quotes in the original) we see in the mirror, or feel with our hands, is still our "externally" observed head or brain. This is why sensations (**s/e**) can share the visual field with other external elements (**s'/e'**) of the head and brain, even in my own case using a mirror. These classes of elements are not identical but should be mapped separately, functionally, side by side, not one "inside" another, and that is how Avenarius and Mach dispose of the introjection problem.

The second error, that of "bilocation," was made well-known by Arthur Lovejoy in his 1930 *Revolt Against Dualism*, in which he attempted to roll back the realistic empiricist views of James and Russell. How can it be, asked Lovejoy, that my perception of, say, a glass on the table is four feet in front of my head and yet, at the same time, in the occipital lobe of my brain? Again the arena is that of a single observer's sensations, but here the criticism is that the patent *non-identity* of the sensations of the glass and the observer's own "head" creates a problem. For example, I can interfere with the sensation of the glass by extending my hand in front of me and, say, blocking my own field of vision but I can also take an electrode and point at a region of my brain, observed in a mirror, and make the sensation of the glass disappear. It would thus appear that the location of the sensation is in both places at once, out there and in my "head," even within the visual field of one person. So we seem to have a prima facie argument for dualism, and this is indeed how Lovejoy interprets it. But Mach (of whom Lovejoy may not have been aware) already anticipated this criticism in his famous "headless body" picture (Mach 1886/1959, p. 19) reproduced in Figure 2.3.

What we have here in the so-called "problem of bilocation" is a simple fallacy of equivocation, sliding from one term to a totally different term in the course of making an argument (Mach 1886/1959, pp. 61–62). Looking at the headless body picture, the subject will *never* be able to observe in the same place both the glass **s/e** and the brain process, which is the sensation of the glass **s** considered as a physical event in my brain **e** under its physical variations. These are one and the same event with two sorts of variations it undergoes simultaneously, not two different, bilocated events. Moreover, the headless body never actually succeeds in seeing the events in his own brain where the glass appears **s/e**, he only ever sees certain *other* events externally correlated with them **s'/e'**, such as the mirror reflection of his own brain. If he actually succeeds in interfering with the sensation cum physical event in his brain, for example while looking in a mirror and conducting the experiment of siphoning off the neural energy with the 1,000 microelectrodes, then the sensation promptly disappears and again there is no bilocation.

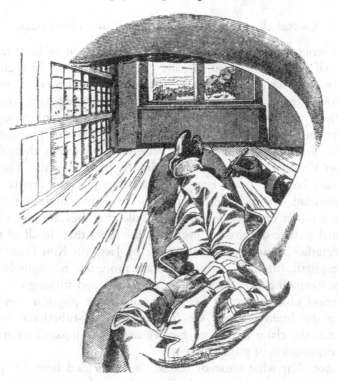

Fig 2.3   Mach's headless body picture

The actual visual field of the headless body "loops around" the observed brain in the mirror, in such a way that the brain tissue may be externally observed in a different place from where our sensations of objects appear, in two separate locations, but the brain tissue *an sich*, where the sensations are actually occurring, is never observed directly in the visual field in the same place as the sensations, so there is never a true case of bilocation at all, contrary to Lovejoy's claim. The headless body can observe his brain in a mirror, but those are just **s'/e'** elements, not the **s/e** elements themselves that he thinks they are. The "x" that marks the location of the sensation in the visual field is matched by a gap in the observed brain in the visual field of the headless body, exactly at the "x" where the brain process identical with the sensation should have been found. It must be, for they are the same event. Mach's construction is thus free of both the spurious bilocation and introjection problems.

Causal closure and mental causation: a first pass

A more serious issue for Mach's view is the question of how sensation energies, if they *are* real and causally effective, show up in their causal influence on other physical events. Is it possible for sensations to play a causal role with each other and in connection to an organism's behavior, interactions with physical objects, and so forth? I'll have more to say about this later, in Chapter 5, but for now I can say that this problem never seriously threatens Mach's position because he identifies the particular sensation **s** with the physical energy in the brain directly. The causal connection between brain energies and other physical events is then straightforward. Many philosophers want to deny that sensations have the causal power to affect behavior and hold that they are just a kind of ineffectual residue on physical events in the brain that do all of the real work. Another argument, made current by Jaegwon Kim (2005), holds that the physical domain, including brain processes, is a causally closed domain. Removing, say, the subject's sensations and sticking only to the third-person observations of his brain, and other physical events, and omitting the brain energies **s/e** from the causal–functional graph in Figure 2.2, the claim is that no further causes are possible or necessary for the explanation of physical events.

It is not clear what sense of "cause" is being used here. Despite the physics-ring to it, there is no such thing as a causal closure law in science. In fact, "cause" is a highly suspect concept that has come in for its fair share of criticism. If "causal closure" is meant to invoke the physical conservation of *energy* for example, we might take Kim to hold that no additional source of energy is needed, or possible, beyond those manifested in third-person observations alone. The subject's actual sensations of green or red contribute no additional energy to the sum total of the universe. This was a common objection to the reality of mental events in the twentieth century, for example, as articulated by Edwin Boring:

> The brain, being material, must be regarded as a closed system: whatever energy is delivered to it must be given up again in motion or lost in heat. There seemed, indeed, to be no way of breaking such a system open to introduce mental links . . . Mental events might parallel its action, but could not cause or be caused by it. (1942, p. 86)

Mach saw this criticism coming and responded as follows:

> I cannot here refrain from expressing my surprise that the principle of the conservation of energy has so often been dragged in in connexion with the

question of whether there is a special psychical agent. On the assumption that energy is constant, the course of physical processes is limited but not necessarily determined with perfect uniqueness. That the principle of conservation of energy is satisfied in all physiological cases merely tells us that the psyche neither uses up work, nor performs it. For all that, the psyche may still be a partly determinant factor. When a philosopher asks the question ... he usually misses the point of the principle of the conservation of energy and the stock reply of the physicist has no intelligible meaning in a case so far removed from the scope of his ideas. (1886/1959, p. 55n)

He is right that this kind of response gets a blank stare even though he is quite correct. The conservation of energy only predicts that the total energy of the universe, or any closed system, does not increase or decrease in any energy change. But it does not predict with any uniqueness what the specific outcome or path of a process may be, or describe the details of the transition. It is not a causal law at all, like Newton's or Schrödinger's law of motion. The principle does not prevent interpolating between the beginning and endpoints of any process, another stage or stages, so long as these do not contribute or use up any net energy. No conservation or closure laws are violated if other internally closed systems exist in which the energy out is equal to the energy in. The concepts of conservation and causation, in the physics sense at least, *are* logically separate.

Mach claimed that the conservation of energy permitted the formation of closed macro-causal "islets" in the overall flux of elements. Like objects, an "islet" would be a function of the elements neither producing nor absorbing net energy. The laws of formation for such macro-causal islets in the causal network are likewise not predicted, or forbidden, by the conservation laws. They simply say nothing at all about them. Nor are laws forbidden which would connect islets or configurations of events to each other, in a kind of higher-order causation, where one configuration either elicits or impedes the formation of a different configuration (Banks 2010). All of that kind of thing going on at the macro-level is compatible with the system of individual events and causal relations. It is possible to imagine that these islets are configurations of elements, following their own laws of formation consistent with lower-order events at an individual level.

The question is, what effect will the formation of these islet systems, and their macro-causal relations, have? This is what I believe Mach hinted at when he said "for all that the soul may still play a partially determinative role." Determinative of what? Well, in the nervous system, perhaps important features for perception and behavior supervene on *overall clusters* of systems of neurons, on the shape and size of configurations rather than directly upon the individual cells and their responses (see Banks 2010). The appearance of certain systems might also elicit or impede the likelihood of the formation of other systems, without any of this contradicting the physical behavior of the individual cells in which these configurations are realized. Hence many of the important features of the nervous system, those which are higher-order and involve pooling of responses, could be found in this kind of system-to-system-level causation. As I have pointed out above, causal maps are especially suited to mapping these kinds of polyadic connections. These views will be examined in more detail later, in Chapter 5. Suffice it to say for now that Mach's causal maps are a remarkably flexible tool for handling the elements in one presentation, and also especially for dealing with polyadic relationships and macro-causal functional relations of all sorts.

Principle of psychophysical parallelism

In the case of elements like the **s/e**, there is just an identity between the sensation and the physical brain energy it manifests, so the parallel between the physical and the mental is satisfied trivially. Moreover, all psychological variations of the elements (trains of association, mental images, even hallucinations) eventually reduce to physiological processes in the brain, which are themselves ultimately part of physics. So again the principle of psychophysical parallelism ultimately reduces to an identity. Mach, however, considered the principle an important guide to empirical research in psychophysics, until such an identity could be established. The question for empirical research is whether such a principle holds for the sensations and the elements of brain tissue **s/e** as they would be presented to a measuring device or an observer studying the same brain process physiologically s'/e'. We are thus dealing with a correlation between **s/e** on one side and s'/e' on the other, and these are indeed non-identical, different events that can be said to run in parallel. What kind of "principle" holds here beyond an ordinary causal–functional relation between different particulars? Mach considered this to be a one-way dependence only (adding the elements in the notation we are using here):

> For every change, difference or variation in the psychical elements (s/e) there must be a change, difference or variation in the physical elements (s'/e'). (Mach 1886/1959, pp. 59–60)

Another way to state it: no mental difference exists without an *observable* physical difference in the associated brain process. Psycho-physicists themselves often point out that the principle is not specific enough in this form, since many variations can be found in the brain as candidates to run in parallel with a given mental variation, not ruling out many spurious ones. Mach of course believed in a thoroughly *physiological* psychology charged with filling in this parallelism of structure exactly. Wherever we find variations among our sensations, there must be parallel variations in brain processes somewhere. But which ones?

At one point, this principle misled him to believe that different sense qualities corresponded to electrical currents in different electrolytes in the axons through which the current was carried, an idea he retracted later. He was also misled by a muscle called the *tensor tympani* near the eardrum, which he thought was the physical parallel of the mental process of "directing one's attention" toward different sounds, when one picks out a single voice in a crowded room for example. The Young–Helmholtz theory of three basic color sensations did indeed lead to the discovery of three cones, but even Helmholtz was misled by the structure of the basiliar membrane in the cochlea, thinking that it was a resonator for standing waves, a thesis disproved by von Békésy (1960) who demonstrated travelling waves instead. The principle of psychophysical parallelism is thus very, very tricky to apply, and leads just as often to error as it does to success.

What we know today about psychophysical parallelism in the case of sensations indicates that specific sensations do sometimes correspond exactly to particular neurons, such as the so-called higher-level neurons which are conditioned to respond to hands or faces, maybe even the Taj Mahal, but only when they are embedded in a particular anatomical *area* of the brain (see Kandel *et al.* 2000). There is no indication that the responses of individual neurons correspond to anything when they are removed from that area. Neurons in the auditory cortex are correlated with sounds, while neurons with a similar internal structure but located in the visual cortex will be correlated with light sensations; but remove them, or switch them, and it is not clear any parallel holds. The quality seems specific to the anatomical area and to the neuron's specific place in the overall network or cluster there, and pattern of firing, not the internal electrochemical nature of the individual neuron or even its individual responses. One will not find colors or proto-colors in the cross-section of an axon, for example, nor the representation of a "hand" inside a hand-detecting neuron. The phenomenal intensity of the sensation, however, does seem to correspond directly with the frequency of the neuron's firing, as many spikes will indeed indicate a more intense sensation. Where is the parallel to be found, then? Between individual sensations and an individual neuron? Transplanting it to another region might change that. Or is it the structural connections between neurons and their firing patterns in groups pooling their responses that correlate more closely with the dimension of sensation quality? Would duplicating the structure of the visual cortex in the auditory cortex cause the neurons there to produce color or light sensations instead of sounds? Or is the parallel still more subtle?

It just does not seem that a principle of parallelism is going to be of much help here in establishing the correspondence between sensory manifolds and areas of the brain beyond bluntly asserting that there is one. And if one succeeds in asserting a parallel, what will really be established is an identity, the mutual vanishing of sensations and certain neural energies in cells, and not a true parallel at all. The empirical research is guided by the principle as a heuristic but only in a general and vacuous sense, since the principle does not come with filling instructions, or a way to distinguish true from spurious psychophysical correlations. Nor is it of much use as a theoretical tool. We cannot even use it to rule out dualism or pre-established harmony, given that the principle is one of parallelism. We only know that the physical or physiological elements fix and determine all of the variations in sensation quality: same physical elements, same sensation qualities, which is trivially true anyway for the **s/e** elements but not obviously true for the functions relating the **s/e** to the **s'/e'** elements. The principle of psychophysical parallelism is not very helpful in filling in these blanks. Indeed, except for the fact that it guides us in tracking down the physiological correlates of mental processes, as we "work our way in" from the **s'/e'** elements to the truly identical **s/e** elements, which would make the principle obsolete if it ever succeeded, one wonders why one cannot do all the theoretical work with psychophysical identity instead.

Mach's achievement

Mach was not a professional philosopher, as he himself admitted rather proudly. Thus he never felt the need to completely firm up his terminology and conceptual framework to prevent misunderstandings. Mach also left several loose ends to be followed up by James and Russell. In particular the detail work in problems of epistemology, the more precise relation of sensations to the brain, and the construction of space, time, and objects out of events, all of which garnered his attention, were still to be dealt with. Nevertheless, Mach remains the first and most significant figure in the movement. He was the first to defend and articulate the position, deriving many of the consequences, answering the main objections, and setting up the framework for an attack on specific problems which would come later. In addition to providing the original breakthrough, identifying sensations with the subject matter of physics, Mach's work is still, I believe, the purest historical exemplar of realistic empiricism.

CHAPTER 3

William James's direct realism: a reconstruction

Introduction: an historico-critical reconstruction

William James and Ernst Mach were close colleagues in psychology, and even met personally in Prague in 1882 to discuss ideas. In 1902, James wrote to Mach to tell him about the development of a new theory:

> I am now trying to build up before my students a sort of elementary description of the construction of the world as built up out of "pure experiences" related to each other in various ways, which are also definite experiences in their turn. There is no logical difficulty in such a description to my mind but the genetic questions concerning it are hard to answer. I wish you could hear how frequently your name gets mentioned and your books referred to. (Thiele 1978, p. 173)

These lectures eventually became James's "radical empiricist" philosophy, which had many features in common with Mach's view in the *Analysis of Sensations*. James certainly started out by building upon Mach's ideas, but he quickly made his own breakthroughs in the theory of knowledge that owe practically nothing to previous authors. In particular, James gave a direct realist account of perception and a causal theory of knowledge which are unique contributions to philosophy. In this chapter, I want to present a reconstructed Jamesian argument for direct realism about the perception of mind-independent objects from the *Essays in Radical Empiricism*, giving what I think is the best historico-conceptual reconstruction of his position.

The operative terms, concepts, and arguments of Jamesian direct realism are probably unfamiliar, even to those well-versed in direct realist theories of perception, and since there are so many different forms of direct realism in the literature, it might help to state some of the main claims of the Jamesian position up front:

1. Taken just as they appear, sensations are real concrete particulars, not mental representations. As they occur, they are neither true nor false of anything, nor do they intentionally represent anything beyond themselves.

2. We distinguish between "acts" or "events" of sensation and "judgments" of perception in which it is asserted that we perceive objects of some kind. Only if judgments are assertible can they be true or false of the objects they are about.

3. Judgments of direct perception of objects, even false ones, can only be asserted in case real mind-independent external objects are present in the environment of the perceiver. This condition is satisfied even when we perceive falsely and mistake one object for another.

4. In a judgment about a direct perception of an object, sensations are linked to external objects through perspectival causal relations, in which case sensations are the *actual proper parts* of external objects perceived directly by us, *not* indirect mental representations.

First, I will show how James distinguished between a perceptual judgment such as "I am now in a room surrounded by real objects in space" and having what James (1977, p. 201) called a "flat" sensation, say of blobs, flashes, and squiggles that merely look like objects in a room but bear no actual resemblance to this experience. On James's analysis, a sensation is simply the collection of colors, blobs, and squiggles it appears to be, representing nothing truly or falsely. When considered in themselves like this, mere sensations, even if they seem quite detailed, are *not* intentional and they lack any intrinsic capacity to represent external objects. Nor, I will claim, do judgments asserted before an array of sensations, in a dream or delusion, qualify as assertions of judgments of perception about objects truly *or* falsely, because they are not in fact judgments of perception at all.

Next, I seek to broaden James's insights by digging deeper into the intellectual conditions of perceptual judgments, showing that they cannot be satisfied within the egocentric perspective of a single subject, but are required to be embedded in an external "perspectival" system of objects and other points of view on them which a single egocentric subject simply cannot occupy all at once, but which are presupposed in a judgment of perception. Here there will be an important connection with Kant's empirical realism. Finally, I will show how I think James has struck a powerful blow against representative theories generally.

James's *Essays in Radical Empiricism*

James began his radical empiricist series of lectures, many published as articles in the *Journal of Philosophy and Scientific Methods*, with a sustained attack on consciousness. The first article in this series from 1904 is provocatively titled "Does 'Consciousness' Exist?" In it, James introduces a neutral stuff called "pure experience," which is the common constituent of minds and physical bodies, and belongs to neither order exclusively. Indeed, these distinctions are only made secondarily in the variations that each bit of pure experience obeys:

> My thesis is that if we start with the supposition that there is only one primal stuff or material of the world, a stuff of which everything is composed, and if we call that stuff "pure experience" then knowing can easily be explained as a particular sort of relation towards one another in which portions of pure experience may enter. The relation itself is part of pure experience, one of the terms becomes the subject or bearer of the knowledge, the knower, the other becomes the object known. (1977, p. 170)

Like Mach, James also denies that the act of conscious representation bestows on the bit of pure experience its concrete quality, or existence, since the same bit would still have its quality and existence as a *physical* occurrence linked to the history of a physical object. The same red patch that we think of as *our* sensation is also a physical brain event tied in with the histories of mind-independent things, so it is not our "seeing of the patch" that makes it red. Removing the act of awareness of the red patch leaves the bit of pure experience neutral, neither mental nor physical. It relates to external objects through its physical variations, and to the mind of the knower through its psychological variations of memory or association. Even calling them "mental" variations as opposed to "physical" ones does not define any fundamental difference, merely a difference of interest for a psychologist or a physicist (James 1977, pp. 136, 193–194). Nor must we stop speaking about the "mind" of the knower, or even about his "acts" and "judgments," just so long as this mind is understood as a collection of its constituent functions or activities.

Are there other bits of pure experience that are *fully* located in external objects, which are not anyone's sensations under any functional interpretation? Here James is far from clear; he may be a neutral monist, or a panpsychist, or even both: see Gale (1999, pp. 198–215) and especially Cooper (2002, ch. 2). Many passages in James do strongly suggest an order of mind-independent "energetic" qualities, as Mach and Russell also assumed, and I see no reason why James would not have done the same (Banks 2003, pp. 144–150, Thiele 1978, pp. 172–176).

James's "two-takings" theory

Most of James's ideas up to this point (the "two orders," the neutral items, the functional ego) are actually found in Mach's 1886 *Analysis of Sensations*, not surprisingly given the close relationship between Mach and James (for more see Banks 2003, pp. 143–151). As I mentioned, however, James finally goes *beyond* Mach in developing a direct realist theory of perception. Mach had only said that he thought the "naïve realism of the common man has a claim to the highest consideration," (Mach 1886/1959, p. 37) and sketched the basic outlines of a causal theory of "knowledge and error." But James does more, giving a précis of his radical empiricist theory of knowledge in "A World of Pure Experience":

> Either the knower and the known are: 1. the self-same piece of pure experience, taken twice over in different contexts; or they are 2. two pieces of actual experience belonging to the same subject, with definite tracts of conjunctive transitional experience between them; or 3. the known is a possible experience either of that subject or another, to which the said conjunctive transitions would lead, if sufficiently prolonged. (James 1977, pp. 199–200)

To ground his theory of knowledge, James lays down a two-takings theory, in which a bit of pure experience is shared between the knower and the physical object, by being simultaneously part of both functional orders, as a point may lie on two different, but intersecting, lines (James 1977, p. 173). In perception, a physical object, or a proper part of it, is directly known by its also being a proper part of the knower's mind when he perceives it, without the intermediary sense-datum or image standing between the knower and the external object, as in the indirect representative theory of perception. The directly perceived object extends all the way from the environment into the mind of the perceiver. This seems to satisfy James's desideratum for a direct realist theory where objects are perceived exactly as they are:

> If the reader will take his own experiences, he will see what I mean. Let him begin with a perceptual experience, the "presentation" so-called of a physical object, his actual field of vision, the room he sits in, with the book he is reading at its center; and let him for the present treat this complex object in the common-sense way as being "really" what it seems to be, namely ... a collection of physical things cut out from an environing world of other

physical things ... Now at the same time it is those self-same things which his mind, as we say, perceives, and the whole of philosophy of perception from Democritus' time downwards has been just one long wrangle over the paradox that what is evidently one reality should be in two places at once, both in outer space and in a person's mind. "Representative" theories of perception avoid the logical paradox, but on the other hand they violate the reader's sense of life, which knows no intervening mental image but seems to see the room and the book immediately as they exist. (James 1977, p. 173)

The pure experience that is a part of the physical object (the book, the room) is also, at the same time, part of the mind that knows it. So James thinks that when I am actually in the room, I perceive the room and the book themselves as they *really* exist, and *not* indirectly through intermediary images or ideas. James gleefully flies in the face of the representative theory of perception, in which external objects somehow cause internal "mental" representations like sense data and what we see are these indirect representations, not the objects themselves or their proper parts. How an external object can cause a mental representation of an entirely different nature, and how the mental representation has the power to intentionally represent external objects are both complete mysteries on the traditional theory.

James takes sensations and images in a realistic, but non-representative, sense. A neutral bit of pure experience can be taken as real merely by taking it to be the complex of colored blobs, squiggles, and flashes that it is. As James puts it: "nothing is there but a flat piece of substantive experience like any other, with no self-transcendency about it and no mystery save the mystery of coming into existence and of being gradually followed by other pieces of substantive experience with conjunctively transitional experiences between" (1977, p. 201). Taken in itself like this, it is neither a physical object, nor is it a sensation. It is just exactly the neutral collection of blobs and flashes it seems to be, and does not represent *any* object truly or falsely because it does not intrinsically represent anything, a point also made by Mach in his discussion of so-called "sense-illusions" (1886/1959, p. 10).

James also points out that there is no intrinsic difference for him between a real fire and a sensation of flames and light, *qua* phenomena, except for the fact that one fire burns real sticks and the other doesn't (1977, p. 181). This fact doesn't make mental phenomena any less real when they are taken as simple occurrences: orange flickering tongues of "mental fire" and "real fire" are on the same footing there. They both exist. Ultimately an experience of orange tongues of mental fire will have further causal links not with burned sticks but with brain energies and other images and associations. But for James this order is just as much a causal order of activity as the "energetic" causal links that link a mental image to an external object (1977, p. 289). Even in the strange world of hallucinations and illogical associations of dreams, there are real facts about mental activity ultimately grounded on the play of central energies in the nervous system. Although the orders may seem radically different, as if following different kinds of laws or connections, we should not be tempted to think of rigidly dualistic orders for physics and psychology. By emphasizing special "mental" variations, James is simply saying that mental life is *also* a realm of natural causality to be taken equally seriously with physics, and ultimately to be merged with it, in a broadened notion of the physical, which of course makes sense given his training as a physiological psychologist:

> Wherever the seat of real causality is, as ultimately known "for true" (in nerve-processes if you will that cause our feelings of activity as well as the movements they seem to prompt), a philosophy of pure experience can consider the real causation as no other nature of thing than that which even in our most erroneous experiences appears to be at work. (1977, p. 289)

In the external, physical order, of course, the blobs and flashes of sensation can be causally linked with the history of an external physical object *and* the human nervous system, but this causal linkage still does not play on any intentional representative relation or intrinsic similarity between sensations and external objects. This point is made by James in his famous "Memorial Hall" example. A bunch of blobs and flashes, even if they look exactly like Memorial Hall, and are shaped exactly like Memorial Hall, will not count as a *perception* of Memorial Hall unless an external causal relation can be established between the blobs and the real Hall:

> Suppose me to be sitting here in my library at Cambridge, at ten minutes' walk from Memorial Hall and to be thinking truly of the latter object. My mind may have before it only the name, or it may have a clear image, or it may have a very dim image of the hall, but such intrinsic differences in the image make no difference in its cognitive function. Certain extrinsic phenomena, special experiences of conjunction, are what impart to the image, be it what it may, its knowing office.

> For instance, if you ask me what hall I mean by my image and I can tell you nothing; or if I fail to point or lead you toward the Harvard Delta; or if being led by you, I am uncertain whether the hall I see be what I had in mind or not, you would rightly deny that I had "meant" that particular hall at all, even though my mental image might to some degree have resembled it.

> The resemblance would count in that case as coincidental merely, for all
> sorts of things of a kind resemble each other without being for that reason
> to take cognizance of each other. (James 1977, pp. 200–201)

Timothy Sprigge (1997) points out that James was explicitly rejecting the
phenomenological tradition of Brentano and his followers, replacing the
"inherent intentionality" of experience insisted upon by the phenomen-
ologists with purely natural causal links which may (or may not) hold
between sensations and an external object. See also Lamberth (1999, p. 78)
for James's critique of intentionality and use of causal links in his
1894 address "The Knowing of Things Together," which appeared later
in the 1909 *Meaning of Truth* as "The Tigers in India." In this essay, James
wonders what it is that really makes a mental image or thought *about* a
tiger in remote India:

> Most men would answer that what we mean by knowing the tigers is having
> them, however absent in body, become in some way present to our
> thought; or that our knowledge of them is known as presence of our
> thought to them. A great mystery is usually made of this peculiar presence
> in absence; and the scholastic philosophy, which is only common sense
> grown pedantic, would explain it as a peculiar kind of existence, called
> intentional inexistence, of the tigers in our mind. (James 1977, p. 155)

He ridicules this, and gives his own causal theory instead:

> The pointing of our thought to the tigers is known simply and solely as a
> procession of mental associates and motor consequences that follow on the
> thought and would lead harmoniously, if followed out, into some ideal or
> real context, or even into the immediate presence of the tigers. It is known
> as our rejection of a jaguar, if that beast were shown us as a tiger; as assent to
> a genuine tiger if so shown ... In all of this there is no self-transcendency in
> our mental images taken by themselves. They are one physical fact; the
> tigers are another; and their pointing to the tigers is a perfectly common-
> place physical relation ... I hope you may agree with me now that in
> representative knowledge there is no special inner mystery, but only an
> outer chain of physical or mental intermediaries connecting thought
> and thing. (1977, pp. 155–156)

Like Mach, who emphasized that causal links may lead to both knowledge
and error, James says that we may reject where the links lead, if they take
us to a jaguar instead of a tiger. Also, even in cases where a mental image
seems to stand as a representative in thought for a distant object like a tiger
in India, James says that the mental image is merely a "substitute" for
sufficiently prolonged experiences of knowledge, or error, which could
be carried out in practice, and does not exhibit any Brentanesque

intentionality on its own without implicitly adding the idea of a *real* causal connection to what is present:

> The towering importance for human life of this kind of knowing lies in the fact that an experience that knows another can figure as its representative, not in any quasi-miraculous "epistemological" sense, but in the definite practical sense of being its substitute in various operations, sometimes physical and sometimes mental, which lead us to its associates and results ... and by letting an ideal term call up its associates systematically, we may be led to a terminus which the corresponding real term would have led to in case we had operated on the real world. (1977, p. 203)

Reconstructing James: the "difference of kind" thesis, "retrospective" and "protensive" certainty

As I am sure James would have said, it would be extremely difficult to isolate raw sensations just as they appear "by themselves," stripped of their connection to other experiences, thoughts, inferences, and the like. In everyday experience we are constantly adding to our sensations with our imaginations and our intellectual judgments, both consciously and unconsciously. Actual sensation is also quite fragmentary and incomplete, so that we often think we sense more than we really do. Thus, it often seems that we could point to the arrangement of blobs and the fact that they are in space and say that they at least *look* like objects in a room. And while it is true that raw sensations do support some simple phenomenological judgments like "this blob is blue and that blob is red," or "this blue one is to the right of that red one," as we could say of painted daubs on a canvas, these judgments fall well short of any perceptual judgment about objects. What do real objects have that a mosaic of sensations cannot simulate? As we shall see in more detail below, objects in space exist from multiple perspectives, not just the single egocentric perspectives in which sensations occur. Objects have unsensed parts like a back, sides, and past and future stages, none of which are present in the mono-perspective of a sensation, which is all surface, all "flat," and all present in a single moment. Sensations also lack any intentional ability to "reach out" to external objects as our perceptions do according to James; sensations are simply individuals with no relations to any other individuals other than causal relations. There is just not enough in raw sensation to even call it a perception of an object.

Let us now consider how best to reconstruct James's theory of perception within a realistic empiricist framework. To do so, we must first address a flaw in what James says above about the "two-takings" theory. The mere fact of the existence of the blobs and flashes as they are, neither true nor false of anything, does not contain *any* direct perceptual knowledge of an external object. So when I am taking the blobs and flashes as real in themselves, I am *not* taking them to be part of an object like Memorial Hall. But when I take them for a perception of Memorial Hall, I am assuming some further *external* causal relation that ties the blobs and flashes to a real building, and they become proper parts of *that* object which I perceive directly. But these further external relations that make the blobs and squiggles a perception of a building are not directly perceived by me. So, where James's theory of perception is direct it is not a theory of perception and where it is a theory of perception, it is not direct.

Let us see if we can resolve this dilemma for James, starting with a phenomenological observation. James claims that in a judgment of perception our thoughts actually seem to "reach out" to the objects in the room around us, a phenomenological feature of perception I think everyone can at least claim to be familiar with: namely that, no matter how many times we tell ourselves that the colors, the room, and the book are only indirect mental representations or pictures in our minds, our perceptions of objects really do *seem* to reach out beyond our minds to the real objects themselves. James claims that this phenomenological feature is *not* an illusion, but a feature marking out a judgment that should simply be taken at face value. Roughly, here is what I think he is saying: "Perception is of objects in space and not of blobs and flashes of sensation, so there need to be objects if there are judgments of perception. Sensations are not part of an object unless there are external causal links, so there are external causal links, which there can only be if there are real objects surrounding me causing my sensations." That inference seems almost absurdly naïve, for could not the same be said of a hallucination of a room? If what James says is true, that we can take the phenomenological experience at face value, he must somehow establish an internal, introspectively knowable difference between the act of *sensation*, i.e., of blobs and squiggles which do not reach out to anything, and the judgment of *perception* of mind-independent objects, such as books and rooms which do exhibit this phenomenology. It is their internally introspected features, in other words, that should serve to distinguish sensation from perception according to James. But, again, how can he possibly make good on this claim when, as everyone knows, sensation and perception seem internally the same from within skeptical scenarios like a dream or hallucination? Why does James not even deem it necessary to discuss skepticism and the dream scenario?

We will take up this challenge in pieces, starting with the internal differences James understands between sensation and perception. This notion that the acts of sensation and judgments of perception are different and distinguishable internally, or introspectively, is sometimes called the "difference of kind" thesis. James shares this thesis with disjunctivist epistemologists such as Hinton (1967). On this view, judgments of perception and acts of sensation are internally different, have different conditions of assertibility and different truth conditions. The sensation is an agglomeration of blobs, flashes, and squiggles, etc. A perception of an object is completely different in its internal phenomenology. I do not see blobs and squiggles, as if painted on a canvas in daubs; I see a world of solid spatio-temporal objects to which my perceptions seem intentionally to reach out. Perceived objects are also very different from sensations. They have sides and a back which I don't sense directly. They have past and future stages which I must imagine adjoined to their present appearance. They exist in other spatial perspectives, in addition to my own egocentric one, all linked together in a systematic way. This chair, appearing perspectivally in a certain way to me, would appear, in a different perspective to someone situated elsewhere, the way an object in space like a penny does from multiple viewpoints. A judgment of perception thus adds further "intellectual" conditions to judgments about the objects being perceived, such as the condition that there is an external object in space in front of me, only some of which is experienced directly through its proper parts that my mind shares with them, and of which my egocentric perspective is only one, embedded in a connected system of perspectives that includes the object from all vantage points. So the perception of the object and the having of sensations of blobs and squiggles are totally different experiences.

But, as the skeptic continues to insist, would not the phenomenology of the room and the objects be exactly as James describes it during a hallucination? Would the subject not still agree that he "sees the objects in the room exactly as they are," and not blobs and flashes, even when all he actually sees *are* blobs and flashes? And would he not extend all of the apparatus of the judgment of perception to his flat sensations anyway, making them seem like objects in a room even when there are none? Acts of sensation may indeed *be* internally different from judgments of perception and the knower may even *know* there is a difference between them, but he may not be in a position to *tell* the difference in a skeptical scenario, and so what does it matter that one is *really* a judgment of perception and the other one isn't. They *seem* the same to us and that should be all that matters to the epistemologist.

I believe that James has already shown us that we do indeed have extra leverage, even from within a skeptical scenario, to naïvely assert that we perceive objects to which our thoughts reach out, though not of course in any mysterious or non-physical way. The key has to do with the conditions of assertibility for making judgments. To assert a judgment of perception presupposes that you are actually *able* to assert it before it can be true or false of anything. Compare the act of speaking a sentence. You not only have to get the sequence of syllables right, you have to know the language and the circumstances under which the sentence is used and probably a great deal more that all has to be correct before your speech can be considered a sentence and not noise. And the same goes for perceptual judgments.

Say I am having a sensation of what seems to me like my being in a room surrounded by books and a table. I may not be able to tell the difference internally between this sensation and a very realistic dream or hallucination. I have no doubt that the purely mental accompaniments of the blobs and the accompanying acts of the mind, even as I infer things about them, would seem to me to be exactly the same in the sensation and in a true perception of the room. Now, if it turns out to be a dream then of course it would not actually have been a perception at all, not even while I was having it; it would have been a sensation of some blobs and squiggles, along with some accompanying mental acts which seemed like perception, which seemed like being in a room, without bearing any external connection to a room, and thus, by James's Memorial Hall argument, bearing no real resemblance to a room at all, now or then while it was going on. In the sensation as it actually was, the "books" had no pages, sides, or back. The sheaf of pages could have been like a swath of white spread out by a paintbrush, as in a Velázquez or Sargent painting, all surface and light. There were no other occupied perspectives on the room besides my own egocentric viewpoint. The interior of the room was like a break-away movie set with no outside, and with just enough filled in to fool the camera. In a classic sense illusion known as the Ames room (Gregory 1994), for example, a wildly oblique array of shapes and surfaces set at angles can seem to resemble a room if one looks at it from one and only one mono-perspective, through a peep hole. Those are the kinds of things sensations actually present at their very best, and perhaps not even this much.

So when the dream or hallucination is revealed to me, say by a kind of instant replay where I can go back and watch the dream again, I immediately deny that I was ever really perceiving anything, and I say that I was only sensing, hallucinating, or dreaming, instead. The phenomenology of the dream on instant replay is that of a tableau of shifting blobs and squiggles now fully recognized as such, along with "flat" mental accompaniments which are not real judgments, any more than syllables are sentences, and on replay the phenomenology of perceptions reaching out to objects is simply *not there* in retrospect, as the whole experience dissolves into surface. Without the presumption of there being an external world of objects around me, the reality of the dream collapses in the replay and I am left with the "flat" phenomena of the dream taken in itself, lacking all representative powers to refer beyond itself. If someday we could replay our dreams back, through miracles of advanced neuroscience, say, I think we would indeed find that they are nothing like the experiences they seem to be while we are asleep.

And with a little thought, I could even say that what is now perfectly clear to me on instant replay was true even before I knew the difference, or knew what I was asserting, or even whether I *could* assert anything. Even when we are unable to tell the difference, there still *is* a difference, a great one in fact, between the act of sensing and making judgments of perception. Just as the blobs and squiggles do not represent Memorial Hall, and never did, so, too, the act of sensing blobs and squiggles, with certain mental accompaniments, is not a judgment of perception, and never was no matter how similar it might have seemed at the time. Seeming to be similar is not the same thing as being similar, as the Memorial Hall example shows. Watching the replay, I am thus able to achieve a kind of "retrospective" backward-looking certainty about the acts performed even under the conditions of a skeptical scenario. These conditions of uncertainty I now see did not actually affect the acts and judgments themselves, which were always internally different, only my ability to tell which I was performing at the time, or which I was unable to perform. The subject's thinking at the time that the acts and judgments were similar, or the same, did not in fact make them so. Seeming to assert a judgment is not the same as actually asserting one, as we can see in the case of asserting the sentence and not just polysyllabic noise.

And with a little more thought, we find that this certainty is not just retrospective, but forward-looking as well. The informed subject knows the difference between the assertibility conditions of the perceptual judgment and the act or event of sensing, even under conditions of skeptical uncertainty. So I can actually *assert* a naïve judgment of perception that I perceive some kind of room with objects in it, just as James insists I can, even under conditions of uncertainty. I assert that I am perceiving a room, when I am, because if I am *not* in fact perceiving a room, but merely some blobs and squiggles, then I know I will not say, later, that I *perceived* a room at all, not even falsely. There are of course cases of false perception where one *object* is taken for another, like mistaking a flat scotch bottle for a book, or a garbage can for a man lurking outside the window, but this dream, hallucination, or total delusion is not a case of false judgment, since the blobs and squiggles will not even support the assertion of a mistaken judgment. In a delusion or dream, instead, I will later withdraw the assertion that I was ever in a condition to perceive *any* objects in the first place. The conditions of assertion of the judgment were not met at all in the skeptical scenario: a perceptual judgment about objects was "asserted" before what was really just an array of blobs and squiggles and their accompanying mental equivalents.

So, finally, we approach that naïve confidence with which James asserts perceptual judgments in the quote above. We can transform our retrospective certainty into what I will call a forward-looking certainty, even from within the skeptical scenario. If I do perceive the room, and the room exists, then my perception goes all the way to the very objects I seem to be perceiving, and in exactly the way I seem to be perceiving them, embedded in an external perspectival system that includes me as an immanent observer, and which fills in all of the other unsensed perspectives besides. If I am only sensing blobs and squiggles, and mental accompaniments of blobs and squiggles, with no real ability to represent a room at all, then I never was in a condition to assert anything about perceiving a room in the first place. All of these facts about future states of affairs are known to me in the skeptical scenario and so I have no difficulty in projecting the judgment forward beyond the conditions of the test and anticipating its eventual outcome, as if I were already watching the replay and forming a judgment about what happened in it. Therefore I *do* have a kind of forward-looking certainty about my perceptual judgments about being in some kind of a room with objects, even if some of my particular judgments about those objects could well be false. At least I am saved from the kind of *total* deception skeptical scenarios offer.

I think this reconstructed argument is true to the way James himself thought, for example in his discussion of the tigers in India. The phenomenal "reaching out" of our perceptions, beyond what he called their "flat" appearance, to objects of some sort is always retrospectively determined by continuing the causal links begun by the subject and seeing what happens, but prospective in the sense that future experience, however it turns out, seems to bestow the phenomenal sense of "reaching out" backward in time on the experiences of the present. This argument is pure James. For if present experience turns out to lack the future connection with the object, then they retrospectively, in the past, also lacked the phenomenology of "reaching out," which was attributed to them by mistake. In both time and space, our judgments of perception make implicit reference to a larger, more all-encompassing framework of spatial and future "perspectives," broader than our single egocentric one, or else they are not really perceptions. Our judgments of perception anticipate and reach out to future temporal and spatial developments along causal links beyond what can actually be given in a present egocentric experience. The judgment of perception already lives in broadened perspectival dimensions in space and in time that extend well beyond the egocentric present, and its only extension in the flat present is like a planar slice through a three-dimensional object.

If this is also what Brentano actually meant by the "intentional inexistence" of objects, not presently given but already perceived as if they were because of some broadened sense of "being" in which our intentional judgments about objects are already embedded, then so much the better, but there is no especially mental or "intentional" mystery, nor is there a need for unexplained powers of representation: all of the conditions cited are natural and causal. There is no need to make a mystery of how our perceptual judgments reach out, when the explanation has to do with the intellectual conditions of judgment and their satisfaction by experience, fully analyzed to include the spatially extended and future causal links which are part and parcel of making those judgments. I think James does here to Brentano what Russell will do to Meinong in "On Denoting," in 1905, taking what looks like "magic" and reducing it in the light of a clear analysis.

Let us now consider an objection or two. Suppose the subject turns the tables on us and asks: "*Why* am I not warranted in asserting I perceive objects, if you admit that the sensation of blobs and squiggles *seemed* to me exactly like a room with books? Are these not the very same internal conditions, gazing on an array of sensations of blobs and squiggles (as you would have it) under which I always *do* assert the judgment that I perceive a room with books? I never directly perceive more than that even when I assert *successful* perceptual judgments." No, James would say, for after you are disabused by watching the instant replay, you, too, would say that the blobs and squiggles do not even "seem" to resemble a room with books and your sensory act doesn't even "seem" remotely similar to a judgment of perception. His response: "But how then can I assert the perception of objects with such naïve confidence when I never really know if I am capable of assertion or not? Suppose that your replay is shown as if it were a dream, and then someone tells me it really happened. All of your squiggles and noise-like mental accompaniments instantly become objects and judgments again!" I believe James would answer that there is no parity and that on the contrary you can *only* naïvely assert this judgment "I am in some kind of room with books," just as it seems to you, since if the room does turn out to be just a collection of blobs and squiggles, you haven't actually asserted anything and no judgments were really advanced. There is no other option *but* to assert a judgment of perception of a room with objects, of some sort anyway, when it seems to you to be so, and that is what you should do in all cases. It is your only play. This would back up James's taking the phenomenology of the room with objects *always* at face value, as he says above.

But what if it isn't incoherent to carry out judgments of perception while looking at an array of what turn out later to be blobs and squiggles in a dream or hallucination? The assertions are well made, they simply come out false. My answer is that there are indeed cases where one perceives falsely: where perceived objects A are mistaken for other objects B. On the Mach–James theory of knowledge and error, the *same* judgments that lead us to correct perception also lead to false perceptions and errors by following out their causal links. For example, a garbage can is mistaken for a man outside the window. But an *object* is perceived in both cases, one truly and another falsely. The judgment that the object is a man is actually assertible of a garbage can too, but false, because causal links can be established that would establish its falsity. The judgment is false because some *other* object is actually perceived, and not because *all* of the objects dissolve into the blobs and squiggles of a hallucination and nothing is perceived because no perceptual judgment was ever assertible of them. These extreme cases of *total* delusion, which are the ones under discussion here, are where we must challenge the idea that the subject could have made any coherent assertion to perceive any object whatever.

The same goes for the mental accompaniments. Take the case of having a dream where you think you are speaking French and pulling it off with panache. But you don't really know French, so the syllables you utter only seem to you to make up a French sentence with the meaning, "Voltaire was Molière's brother." You might wake up and realize this is a false assertion, when you correctly phrase an English sentence, expressing a proposition that *can* be true or false, but did you succeed in making the assertion *then* by uttering a string of syllables and calling it a French sentence expressing the proposition that Voltaire was Molière's brother? No, it never happened, no matter what you believed at the time. This so-called "assertion" of a proposition wasn't anything beyond an agglomeration of syllables and an accompanying feeling that one was making sensible assertions. The feelings and sensations certainly exist in themselves and are neither true nor false; they should also be taken in the direct sense as exactly the strings of syllables and feelings they seem to be but without further significance beyond themselves.

These sorts of arguments, which are rightly associated with Putnam (1981) and the later Wittgenstein, have their roots in James's Memorial Hall example, completely independent of the philosophy of language and issues about meaning or reference, and they are better developed in the way I am suggesting, first to establish a retrospective certainty and then finally a forward-looking certainty from within the skeptical scenario. Putnam recognizes some kind of retrospective certainty about hallucinations as a possible reading of James (Putnam 1990, pp. 248–249) but rejects it since he thinks it makes James an anti-realist about past experience. Putnam seems to think the perceiver who has hallucinated later has to say that he never experienced anything, when clearly he did at least experience the blobs and squiggles of his sensation, which really do exist as phenomena in their own right. But James is only saying that experiences that retrospectively do not measure up as perceptions still existed in the past, exactly as they are were in themselves: they are facts like any other. Also according to Putnam, James never refuted the skeptic (p. 246) but simply held up direct realism as a viable alternative to the "sense data" theory that we never perceive external objects directly (p. 251). James may not argue in the traditional *way* against the Cartesian skeptic, but he still shows us much more than an alternative to sense data.

To sum up the position we have reached so far: James's quote above, about taking the room and its objects to be directly real perceptions of those external objects "as they exist," now makes more sense, where it seemed hopelessly naïve before. On the Jamesian view, the perceptual judgment that I perceive parts of real external objects, in some way, is always asserted *full strength* or it is not asserted at all. I do not, for example, assert a watered-down disjunction: "*either* I am perceiving objects *or* I am merely sensing my own blobs and squiggles." Thus, when I do perceive, my thoughts go straight through without intermediary to the external physical objects they seem to be about, and of which my sensations (colors and all) are their true proper parts. Even false judgments presume some general conditions like this, even if we take some objects for others. Putting it all together in argument form:

1. Either I am in some kind of real room with objects arranged around me, or I am before an array of my own sensations of blobs and squiggles that do not even resemble any room with objects.

2. If I assert the judgment that "I am perceiving some kind of room with some objects around me" and I am really before an internal array of blobs and squiggles, then I will never have succeeded in asserting a judgment to perceive anything.

3. If I assert the judgment that "I am perceiving some kind of room with objects around me" and I am really in a room with objects around me, then I will have succeeded in making the assertion and the assertion will have been true at least of some general sorts of objects, though it may still be false of specific objects.

4. I know (1,2,3) under the conditions of a skeptical scenario.

---

5. I know I can always correctly assert, with forward-looking certainty, the naïve judgment that "I am in some kind of a room with real objects around me" when I am having that experience.

Intellectual conditions of perception

As I see it, the remaining problem with the "no common kind" thesis is that we are owed some explanation of why the acts of sensation and judgments of perception still *seem* the same in skeptical scenarios, if they are really so different that no one would take one for the other when shown the instant replay. Acts of sensation and judgments of perception have to be both absolutely indistinguishable in skeptical scenarios, and yet retrospectively recognizable as internally different, but how can this be? Certainly the purely sensory components of both acts and judgments are completely the same: they both consist of blobs and squiggles and mental accompaniments, and there is no internal way to tell apart the blobs and squiggles of a sensation from the directly perceived proper parts of external objects in a perception. But there always is a difference, and there will have been a difference, between acts and judgments we perform even under conditions of uncertainty. So, since perceptions and sensations *are* internally different acts, and since the difference does not consist in their sensory contents, it must consist in something else: namely the intellectual contents of a perceptual judgment. It is therefore true, what the disjunctivists and Jamesians assert about the difference in kind, but, to make a Kantian point, it is rather intellectual thought and understanding somehow that contribute this difference to the perceptual judgment, which is not contained in the mere act of sensing. So, for example, on replaying the dream, we consciously withhold these intellectual features of perceptual judgment and so the phenomenological feature of perceptions "reaching out to objects" disappears, replaced by the sensation of a flat tableau of shifting blobs and squiggles. If one thinks the instant replay would still look like a real experience, one must really consider what it would be like to observe the instant replay of a dream or hallucination while withholding *any* of the accompanying intellectual judgments we are always advancing in perception. Since we have already assumed that there is no "inherent intentionality" to sensory contents alone, the phenomenon James refers to, of experiencing perceptions that actually "reach out" to external objects, seems to be effected purely by the intellectual contents of the judgment of perception which we are nearly always extending when we are convinced of the reality of our experience of objects, and which we must now investigate.

I have been suggesting that a perceptual judgment like "I am in some sort of room with objects" has the following preconditions: my egocentric experience of "walking around a room viewing various objects" from my own mono-perspectival point of view cannot be all there is. There must be, at a minimum, an occupied system of perspectives surrounding these objects and me, which includes my own immanent mono-perspective as one, but most of which I am *not* able to observe, a system which fills in all of the possible points of view on the room from those other perspectives I do not, and cannot, occupy. In those other perspectives, chairs and books will have backs and sides unseen by me and a history in time beyond the present in temporal perspective. The room will not be some Ames room or breakaway movie set with just enough walls and furniture to fool the egocentric viewer. A town seen out of the window will not be a Potemkin village with fake storefronts and cut-out windows, with false interiors and simulated figures walking back and forth. In a perception, the room and its objects will exist from *every* perspectival point in it, which they clearly do not in these nightmarish mono-perspectives.

It is literally impossible for any given experience of objects-to-a-subject to internally represent or simulate a real object, and its entire sum of circumambient perspectives, because no one can literally occupy more than one point of view at once. These missing perspectives occupied by objects must therefore be filled in by the intellect and the imagination, but they must also be *really* occupied by objects on the side of the external environment as well in order for judgments asserted of them to make sense. Thus in order for my intellectual judgments of perception to even be assertible, I must be surrounded by a perspectival system of objects of which my mono-perspective is one, and where any perspective can serve as a vantage point from which to represent all of the others.[1] These assertibility conditions are known to me internally, of course, and I can argue, as above, that if I do turn out to occupy a mono-perspective that is all surface with no other occupied perspectives, and my judgments filling in missing spatial and temporal perspectives were only the mental accompaniments of judgments, my additions only imaginary and flat, I will then say that a judgment of perception was not asserted under those circumstances and withdraw it no matter how realistic it might have seemed to me at the time.

Also to be eliminated from consideration are situations involving "deviant" causal chains to external objects that are not even possibly interpretable as perspectival relations between a perceiving subject in the same subject–object system of perspectives as his perceived objects. The system of perspectives must embed the subject's immanent, internal viewpoint as one such perspective within it. It must be a systematic and connected continuation of the subject's internal perspective outward. It cannot be replaced by some *other* set of perspectives of objects of which he is not aware, nor can he be in contact with them in some indirect, non-standard, non-perspectival way which he himself would not assert when he talks about "these objects he perceives with his eyes as being five feet in front of him." If, for example, an evil scientist wires up the subject's brain to a simulation, but takes care to connect the sensations of the simulation via deviant causal connections to *other* external objects in a secret room, in *another* isomorphic system of perspectives that exactly duplicates those of the simulation, the subject still cannot assert any perceptual relation to *those* objects. *His* own occupied perspectives are still empty and he really just occupies a mono-perspective as before.

A clever philosopher placed in this situation might try to assert that a scientist is connecting his perception of a room to *another* perspectival

---

[1] See Friedman 2012 for an exegesis of this "perspectival principle" and the role it plays for Kant. See also Brewer 2002 for the view that this "perspectival" content of space and time is directly perceived.

system of objects in *another* room, and make guesses about *those* objects, and about the scientist's occupied perspective too. But these missing perspectives on the *perceived* room are still not occupied: some *other* system of objects, around some *other* room, is occupied and the missing perspectives on the *actually perceived* room remain empty. The philosopher may *think* whatever he wishes, and he may even guess correctly, but he must give up the idea that he *perceives* the room he guesses about. Perception is still, for all its intellectual pre-conditions, a sensory matter of being in direct contact with objects and their proper parts in a way that is projected outward from the perceiver's own perspective in the way he intends it, not through indirect or merely representative contact through *any* imaginable links someone else might think up. If I am not linked to the object in the way I perceive myself to be linked to it, then it isn't really my perception at all.

Other perspectival systems: Kantian "empirical realism"

What this argument does not show is what *sort* of perspectival system of objects surrounds the observer when he makes these perceptual judgments. What are the intellectual conditions for a valid perspectival system continuing a subject's own perspective and perceptions outward to an environment of objects? Space and time are certainly one way to satisfy the intellectual requirements of a system of perspectives. Any point within space can be chosen as the origin of coordinates for representing any other point elsewhere in the same space; any point in time can serve to represent any other point within the same time line, as a past, present, or future point in temporal perspective.

To introduce a key Kantian observation, an objective intellectual "skeleton" of a subject–object perspectival system undergirds the sensory space-time form of our perceptions, as Kant declares at several key points in the transcendental deduction, particularly in the B-edition (but see also Kant 1787/1998, A 107–109). In brief, the argument is something like this. Kant declares that the "synthetic unity of apperception" is the highest principle guiding the understanding in the construction of experience, higher even than the categories, for it is the synthetic unity that ultimately justifies the categories' application to any possible experience (this is what transcendental "deduction" means). The synthetic unity takes all intuitions and sensations delivered by sensibility as a disordered bundle of "snapshots," if you will, and unifies them under the principle that they are one and all experiences of objects-to-a-subject.[2] When I walk around a house taking in the different sides in a certain order, this ordering and the egocentric perspectives of the snapshots are determined by the series of vantage points taken up by a subject (including the ordering of his inner states in time) to an object in a distinct sequence of perspectival views of the house. If these experiences were presented in a series of disordered photos of the house lying in a heap, it would still be possible to intellectually reconstruct the objective series via the categories under the overall heuristic principle that they must present an ordered series of perspectival views of the object to an observing subject.

In the B-deduction (B 151–152), Kant distinguishes between an "intellectual synthesis" of experience, carried out by the categories and the synthetic unity and applicable to other forms of sensibility besides ours, and a "figurative synthesis" carried out by the imagination and sensibility under the direction of the understanding. For Kant, the construction of the underlying perspectival system of vantage points of subjects and objects, the "intellectual synthesis" of the understanding, is what is really objective about spatio-temporal representation, not the schematized space-time form contributed by sensibility and the imagination, which actually depends on the former for its objectivity. This intellectual synthesis of a system of perspectives would apply even where the intuitions of space and time were replaced by other forms of intuition or sensibility. And when spatial and temporal intuitions are present it is the intellectual synthesis again that explains why synthesized constructions in space and time, like drawing a line or synthesizing a certain number, are actually objective, as in math and physics, and are not just arbitrary constructions of the human imagination with no application to experience. For example, where we adjoin images of previous and future stages of the construction to a presently sensed stage in counting, drawing out an object like a house, or tracing the parabola of a falling body, these constructions are actually objective, empirically real constructions.

Kant also shows in two examples (B 161–162) that it is the understanding that actually directs the sensibility and the imagination in these constructions of objects, such as taking in the sides of a house and watching water freeze. He says we can "abstract from" the spatial and temporal form of these experiences and consider ordering the states intellectually, with the

---

[2] The principle is metaphysical, not psychological. *Any* reality is capable of subject–object representation by taking up a standpoint within it and representing the rest from that vantage point, with all other internal standpoints equally valid.

categories and the synthetic unity of apperception operating on receptive sensory and intuitive content of whatever nature. The intellectual synthesis of ordered perspectival views of objects-to-a-subject would apply even if the sensory, spatio-temporal form of the experience were different, perhaps for other thinking beings with differently constructed sensibilities.

In fact, there are many ways to consider constructing representation systems that have the same intellectual "hard core" of a perspectival system of objects-to-a-subject, not just spatio-temporal ones. Leibniz's system of monads is an example of which Kant was particularly aware, and we can readily think of others, like the abstract, infinite-dimensional Hilbert space which we use to represent physical systems in quantum theory. Or consider, as we will below, constructions of objects out of elementary "point-events" in abstract quality or property "spaces" which still retain a perspectival structure but lack extension, affine structure, and even a metric (see Banks 2013c). These other possible systems count as candidate environments for my perception. We perceive objects intellectually, objects which we "reach out to" through perceptual judgments, and not through our sensations alone. Those judgments have a firmer intellectual "hard core" of assertibility and truth conditions that the external world can satisfy in many different ways. Even though we directly perceive the proper parts of mind-independent objects, then, we cannot be said to perceive more than their overall intellectual structure in an external perspectival system in their relation to us: the structure of objects-to-a-subject which may be realized in many ways.

To sum up this Kantian line of argument, we perceive the proper parts of empirically real objects directly but not *uniquely*; only the intellectual content of our perceptual judgments represents to us the external objects our perceptions are connected to. In the sense that spatio-temporal objects are one such way to satisfy these conditions, then I do indeed *directly perceive* an environment of spatio-temporal objects around me in a room and I perceive their proper parts directly. But other rational beings could reinterpret my spatio-temporal perceptions in terms of their own empirical representation system and we would both be equally right about those objects, since whatever the other beings imagine, they share the same perspectival structure of objects-to-a-subject, which I represent to myself in a spatio-temporal way, but the other beings represent some other way with the same underlying structure. I can always translate their system into mine and vice versa by sticking to the common structure.

This idea that we perceive spatio-temporal objects directly, but not uniquely, is what I believe Kant really meant by calling space and time

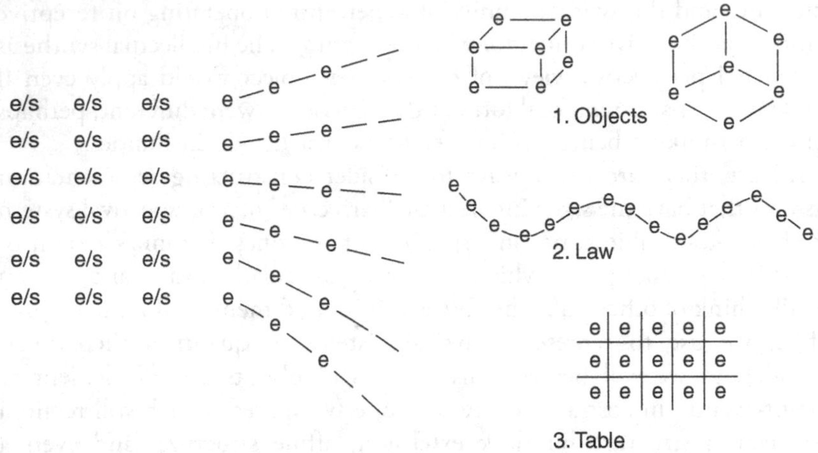

Fig 3.1   Alternative perspectival schemes over same e-elements

"empirically real" but also "transcendentally ideal," since spatio-temporal representation of objects cannot be applied beyond human sensibility but the intellectual preconditions of spatio-temporal representation can. Spatio-temporal objects in perspectives always serve us as an objective empirical representation, as in physics, but this representation is only *one valid way* to represent many possible intellectual systems of perspectives, of which space and time are only one. Other rational beings would thus agree on the *intellectual conditions* for asserting a perceptual judgment, but not on the sensory spatio-temporal form we might give to objects (consider Figure 3.1). The mind-independent elements, **e**'s, can be ordered several different ways, exploiting their various functional connections to each other to make up a perspectival system of relations around the subject, not just into mind-external spatio-temporal objects, or mechanisms, as in choice 1, but also into the forms of abstract laws and tables, and much else besides.

We do, however, perceive objects using our form of sensibility, and yet we *are* justified in taking this admittedly subjective form of representation of objects in a completely realistic and direct sense, just as long as we stick to the hard-core intellectual features of our representation: the perspectival structure of objects-to-a-subject, however that may be realized.

Is this anti-realism? I think not. Our scientific theories can involve realistic spatio-temporal models and mechanisms of natural events and processes continuous with our own form of spatio-temporal perception, and we can even think of these models in a realistic sense so long as we do not thereby limit ourselves *only* to this form of representation or attribute what may only be sensory or visual aspects of the models and mechanisms to reality *an sich*. The intellectual content of these representations is the guide to objectivity. So long as we are aware of this, we are free to pursue whatever models and analogies we wish.

Is there also an historical connection between James's direct realism and Kant's empirical realism? Like Kant, who struggled to explain to literal-minded readers how he could be empirically realistic about spatio-temporal objects, yet insist that objects need not be spatio-temporal in themselves, James, too, was at pains to explain how he could be pluralistic *and* direct in his realism all at the same time. In an oft-quoted letter of August 7, 1907 to Dickinson S. Miller (reprinted in James 1920), James gives a homely example of beans to illustrate this. One perceiver classifies the beans by weight, another by color or size. All have legitimate and empirically real systems of classification determined by the subject's interests; they are all direct and real, but no single classification has a unique monopoly on the one true system of beans *an sich*, nor do they even really disagree.

Is there an historical connection here? Perhaps. James was teaching Kant before he began the radical empiricist essays, from 1897 to 1899, although his remarks there are surprisingly dismissive (James 1975, Carlson 1997, pp. 363–364). As Carlson points out (p. 364), James was irked by colleagues who remarked on his similarities with Kant. Apparently Hugo Münster-berg told James that there was nothing in his philosophy which could not already be found in Kant, and accused James of being ignorant of this. I am therefore not sure James ever understood Kant's empirical realism.

Epilogue: against representational theories

One consequence of the view presented here is that mental representations so-called are *never* complete in themselves if they are going to represent external objects. If we take them in the completed sense, in themselves, we are left with non-representative blobs and squiggles, or nightmarish mono-perspectival dreamscapes, that do not represent anything. If an experience is going to represent anything, it must be *incomplete*, needing to be filled in, for example, by the intellectual contents of a judgment that makes the experience "reach out" to external objects. But these objects cannot appear *in* experience, entire and complete in themselves, else the experience would *already* be complete, lacking nothing, and so would not reach out to anything either. The lesson to draw is this: intentional judgments and perceptions must be *incomplete* fragments of an unseen, but connected and presupposed whole of experience-reality in order to "intentionally" represent anything. They cannot be an already internally complete simulacrum of the world à la Descartes, for if mental phenomena are complete, then they represent nothing beyond themselves and have no "intrinsic" intentionality at all. And if they do "intentionally" represent external objects, then they are *not* complete and the missing objects and perspectives which do complete them must be anticipated by an intellectual judgment of perception. "Intentionality" so called is therefore not a fundamental phenomenon at all, as Brentano thought, but a relative condition of incompleteness. When multi-perspectival intellectual judgments are embedded in a limited perspective of one subject, they appear to be referring intentionally beyond themselves, but when the perspective is broadened we find only a homogeneous multi-perspectival causal network of links and the "phenomenon" of intentionality simply vanishes. It was an artifact that only existed when we took an artificially narrow and egocentric view of perception.

Descartes thought a picture, which is a completed representation or arrangement of objects, always required something *else* for it to be a picture of, but this view is also wrong, for two reasons. One, the picture is a completely existing thing which requires nothing else to complete it; it has its own complete "formal reality" as Descartes himself says. But when your representation has all the structure or "formal reality" the object has, why does it need the object? What would the object add to an already complete picture? Two, the external reality that supposedly *does* complete the picture is no more "formally" complete than the picture was originally. We can add a real arrangement of real buildings and real sailboats around a bay to the painted ones in a picture, but each is a completed thing on its own that does not intrinsically represent the other. Adding the external reality to the picture merely adds a picture to a picture. Neither arrangement *lacks* anything for which it needs the other. If they are not causally related, then why should they be related to each other at all? Isomorphism of structure does not, in any way, demand a further representative relation between those structures. This was James's original point about Memorial Hall: let it have as perfect a similarity to the Hall as you like, but without an external causal relation there is nothing to paste them together and certainly no magical "representative" relations to suggest one is intrinsically related to the other, a relation which may be accidental or arbitrary after all (see also Putnam 1981 for his effective critique of "magical" theories of reference and ensuing skeptical consequences for model theory in logic).

James's Memorial Hall argument thus deals a powerful *coup de main* not only to theories of mental representation but also against *any* theories of representation that work by mere similarity or isomorphism of structure (a point taken up by Wittgenstein and Putnam). As an afterthought, take it for what you will, I think our having essentially "incomplete" mental experiences, needing to be completed by intellectual judgments about their real external relations to mind-independent objects, makes a great deal of evolutionary sense, especially as a space-saver in our skulls. Why duplicate effort modeling the whole world in a "simulacrum mind" with its own complete internal world representation? Why not just sketch in some incomplete fragments, model only as much of an "interface mind" as it takes to directly perceive limited bits of the world, and as for the rest project intellectual judgments that reach out transcendentally beyond these bits, letting the world fill in the rest?

CHAPTER 4

Russell's neutral monism: 1919–1927

Introduction: a continuity view of Russell's development

Bertrand Russell's neutral monism developed in a period from about 1914 to 1927 and, according to Russell himself, remained his philosophical view from that point onward. In a personal communication to Elizabeth Eames (1967) Russell said: "I am conscious of no major change in my opinions since the adoption of neutral monism." Many philosophers and historians of philosophy seem to lose interest in Russell at about this time in his career, feeling somehow that he was no longer doing good work, or that he was becoming despondent about philosophy, as he wrote in a letter to Ottoline Morrell in 1916. Commentators have repeatedly sought to marginalize neutral monism as a brief period in Russell's thinking, after "On Denoting" and *Principia Mathematica*, and before his so-called conversion to commonsense "scientific realism" in the *Analysis of Matter* (1927).

The record on neutral monism has since become clearer, in important work by Robert E. Tully (1993, 1999) and Michael Lockwood (1981). According to Tully and Lockwood, Russell should never have been seen as a "phenomenalist," even in earlier writings like the *Problems of Philosophy* (1912). Nor should we see any fundamental change of view "from neutral monism to scientific realism" occurring in the period between the *Analysis of Mind* in 1921 and the *Analysis of Matter* in 1927, rather just the continuous evolution of his neutral monism into a realistic position compatible with physics and psychology.

We also understand more about the various stages through which Russell arrived at his version of neutral monism (Tully 1993, Hatfield 2002). As Tully has shown, Russell abandoned his theory of acquaintance, starting with "On Propositions" (1919), because of specific difficulties in formulating the concept of knowledge as a "multiple relation" in the manuscript *Theory of Knowledge* (Russell 1914/1984). This was not the only reason Russell gave for his conversion, of course. He also says that he realized the act of acquaintance could not be discovered by introspection (see Tully 1999, pp. 351–2), as Mach and James had urged, and that neutral monism could unify the data of physics and psychology in a more parsimonious way than the assumption of two categories of things, the mental and physical (Russell 1959/1997, pp. 103–4). But since Russell knew these arguments already in 1914, and discounted them, I believe with Tully (1993, p. 20) that "the deeper story" does have to do with that *Theory of Knowledge* manuscript and issues touched upon therein.

Finally, I believe we can identify a realistic and naturalistic bent to Russell's thinking which pushed him away from abstract philosophy of language and toward realism in philosophy more continuous with science, including psychological realism (as Tully 1993 and Hatfield 2002 have already said). Some of this was successful and some of it wasn't. Russell's psychologically realistic "image propositions," and the representative propositional theory of knowing he eventually based upon them (in "On Propositions" and the *Analysis of Mind*) were strange and dualistic compared to Mach's and James's more streamlined causal theory of knowledge *and* error, which did not demand propositions or representative relations over and above causal relations, and more importantly, which had no difficulty with the representations of false facts, since for them error is just a form of causal non-agreement, an issue that plagued Russell at this time. Russell continued to hold to an essentially representative view of knowledge either via multiple relations and complexes, or via image propositions, and never embraced a truly causal theory of knowledge and error à la Mach and James.

However, Russell's neutral monism about the events of physics and psychology, his physical theory of perspectives, and his attempt to connect his logical constructions of events with the actual physics of relativity and quanta in the *Analysis of Matter*—all of this was certainly a positive development of his thinking away from linguistic philosophy and toward naturalism, even if the movement was not quite complete. I conclude that there is a cleft separating Mach and James's pure neutral monism from Russell's on the theory of knowledge, which led to a lingering dualism between psychology and physics in Russell that is not there in Mach and James; but there is much agreement elsewhere, especially on realism.

The theory of acquaintance and "sense data"

In 1912, when he wrote the *Problems of Philosophy*, Russell held his theory of acquaintance. Acquaintance, according to Russell, was *the* fundamental act of the mind established between the mind of the knower and the immediate "sense data" present to him (1912, pp. 38–39). Russell also believed, for a time, that the mind was directly acquainted with universals, properties, relations, in short whatever was needed in order to construct *propositions* about the world capable of being true or false of external states of affairs. Given a stock of sense data, and these other notions, the knower could construct propositions even about things with which he was *not* directly acquainted, constructing the rest by definite description. We thus extend our knowledge outward to what we express indirectly through true or false propositions which represent external states of affairs to us. James had made an earlier distinction between knowledge by "acquaintance" and what he called "knowledge about" (James 1977, pp. 199–200), where the latter is connected to the former by a series of indirect and even potential *causal links* to what James calls their "terminus." But despite these terminological similarities with Russell, James clearly should *not* be considered a follower of the theory of propositions, intentional mental representations, or irreducible mental acts of acquaintance, despite having coined the term.

In a Russellian proposition of this period, the components are the very objects, properties, and relations of the world as they are present before the mind, not representations. Russell's theory of propositions in *Problems of Philosophy* is therefore sometimes known as the "Fido"–Fido theory, because it is Fido, or the sense data of Fido, with which we are acquainted, and not his sense or individual concept or mental representative, that is a constituent of all of the propositions about him. The proposition, though it may only be a description of something not immediately present, is at least made up of real things in contact with the mind, not mental images or representations, and certainly not abstract Platonic senses as in the rival theory of Frege, which Russell rejected.

Russell had already shown how to handle non-referring singular terms like the "Present King of France" in the 1905 "On Denoting" by allowing the knower to be acquainted with all of the parts of a proposition, like the property of being King, the country France, and yet substituting the variable x in a definite description for a name, so that one could coherently deny the existence of the present King of France without denying the existence of these properties, or assuming acquaintance with a non-existent "subsistent" object as the referent of the name. By contrast, a Fregean proposition (*Gedanke*) was a creature of sense, an abstract, mind-independent object made up of the abstract concepts, or hypostasized *meanings* of names, predicates, and relations. Frege's theory required a three-way connection between our mental images and thoughts, the abstract meanings and propositions our words and thoughts express, and their references to the True and the False. Frege's theory distinguished meaning from reference, and could explain indirect discourse and belief contexts by allowing terms in those contexts, called "that-clauses," to refer to their customary senses not their usual references to objects. In the sentence "Mary believes that Paris is the capital of France," the reference of the that-clause, according to Frege, is to the customary senses of Paris, capital of, and France combined in the proposition that Mary believes, not the actual city of Paris and the country of France, which is of course quite odd, since most people probably would not say that the reference of their beliefs is to these abstract concepts, but to real objects and relations.

Russell could well appreciate the virtues of senses as explanatory devices for a theory but he could not see how such things could *exist* in the real world, albeit in a Fregean third realm. He even called Fregean propositions "unrealities" and "fictions" (Tully 1993, p. 200). Instead, Russell says that universals "subsist" like numbers but do not really "exist" (Russell 1912, p. 76, and end of ch. 9). Russell's subsequent views of the existence (or subsistence) of universals and mathematical objects became increasingly skeptical. To explain why, many commentators point to Wittgenstein's critique of tautologies as empty, and to Russell's later view that mathematical objects like numbers only have a kind of what-if "conditional existence." Russell's enthusiasm for abstract objects waned considerably as time went on.

In his *Problems of Philosophy*, Russell called "sense data" the immediate givens to the mind: colors, sounds, and other experiences whose existence could not be doubted. He emphasized, however, that we are only aware of sense data and not the external objects they seem to represent like pennies and tables (see Russell 1912, ch. 3). These are made of atoms and molecules, which differ from the sensed qualities of the datum. However, even though the sense datum is not a piece of the mind-external table, it *is* still a real proper part of the "physical" world in its own right, probably an event in the brain. Russell says in "The Relation of Sense Data to Physics" (Russell 1914/1957, p. 144): "I regard sense data as not mental and as being, in fact, part of the actual subject matter of physics." Russell uses the word "physics" here, I think, to mean an enhanced view of physics which would include the events of psychology within it, such as sensed colors or sounds, as legitimate "physical" events. I do not think he actually means to say that current "physics," in the narrower sense of objects in space and time, includes items like color sensations. He also emphasizes later in the

Analysis of Mind (1921, p. 10) that the neutral data are *neither* mind *nor* matter in the customary sense of those terms and so do not belong to physics *or* psychology as traditionally and dualistically conceived. Whenever he says that he considers sensations to be "physical" or the "subject matter of physics" (see above), I therefore think we should take him to mean "physics" in the sense of this new broadened enhanced physicalism of neutral monism that would include psychology within physics. He is not for example suggesting that physics as it is today explains our actual sensations of color or sound, but that an enhanced physics would include these phenomena alongside other physical events and objects.

"Sensibilia"

In 1914, Russell published a series of articles in the *Monist* called "On the Nature of Acquaintance, I, II, III," which were part of a manuscript called the *Theory of Knowledge* (Russell 1914/1984). These articles actually began with an *attack* on neutral monism, or "the theory of Mach and James." Russell said that he agreed with them that sensations *could* be taken to be non-mental parts of the physical world and that the recognition of this fact was "a service to philosophy," to see that "what is experienced may itself be part of the physical world and often is so" (p. 31) or that "constituents of the physical world can be immediately present to me" (p. 22). Russell did *not*, however, agree with Mach and James that the difference between sensations and physical objects was merely one of ordering or functional relations. The relation of acquaintance, for Russell, still made some kind of basic difference between a "sense datum" and what he called "sensibilia." In "The Relation of Sense Data to Physics" he put the difference this way:

> I shall give the name *sensibilia* to those objects which have the same metaphysical and physical status as sense-data, without necessarily being data to any mind. Thus the relation of a *sensibile* to a sense datum is like that of a man to a husband: a man becomes a husband by entering into the relation of marriage and similarly a *sensibile* becomes a sense datum by entering into the relation of acquaintance. (Russell 1914/1957, section III)

The sense datum is still being distinguished from the *sensibile* by entering into this relation, although the same particular which we call a sensation of red is also a "physical" particular, or sensibile, in some enhanced sense of physical, similar to what Mach meant by a physical "element" or James by "a bit of pure experience" in its physical variations. The term "sensibilia" *also* included purely physical events, or interactions, making up mind-independent objects that could never be considered *anyone's* sensations. Tully says, for example:

> Besides actual data, other particulars assumed to be qualitatively similar to them were also claimed to be constitutive of external objects ... Since such classes of particulars were thought by him to be both objective and real, it is evident that his sensibilia—despite the term's etymology and its association with phenomenalism—are not merely possible constituents of one's experience; they exist independently and need not ever come within the ambit of sensory experience. (1993, pp. 8–9)

For example, in *Our Knowledge of the External World* (1913/1926), when Russell expands the physical object, like the star Sirius (1921, pp. 101–102), into events which represent all of its perspectival effects on, or causal interactions with its environment, only some of those effects will be in interactions with human brains and result in those sensibilia which could be considered sense data when they enter into the relation of acquaintance. There are other sensibilia which *never* enter into a causal relation with human observers but which Russell clearly believed existed in order to fill in the unobserved perspectives of an object, such as events we photograph or measure with other devices as the object causally interacts with them. These interactions are real concrete events in nature in which "what common sense regards as the effects" (ibid.) of objects are manifested.

After giving up on the "sense data/sensibilia" distinction, both of these terms disappeared and Russell spoke instead (in the *Analysis of Matter*) of "percepts" and "event particulars," where by the latter he meant completely mind-independent events, unless of course they are event particulars which occur in our brains and then these event particulars can also be called percepts as well as "physical" events, in the broadened sense of the "physical" I mentioned above, which includes sensation (see the discussion in chapter 37 of *Analysis of Matter*, esp. p. 384; see also Lockwood 1981, 1989 for the idea that the notion of a particular "event" is the most all-inclusive category, including the percept as just an event particular that occurs in a human brain).

So to recap, there are actually three sorts of "particulars" in this period of the theory of acquaintance: *sense data*, which stand in the relation of acquaintance to the mind; *sensibilia*, which are the same particulars we were acquainted with as sense data, but which are considered under a different physical aspect to be part of "the subject matter of physics"; and sensibilia with which we cannot be acquainted at all, making up the external objects and their interactions with measuring devices and each other. Russell eventually came to think that these last mind-independent event particulars had quality or what he calls "intrinsic character" like our sensations. Of that, more below, when we consider Russell's *Analysis of Matter*.

Giving up the theory of acquaintance (1919–1921)

So until 1919, acquaintance was the basis for distinguishing the sense data from sensibilia. But it was also the important foundation of the project Russell was working on in the manuscript *Theory of Knowledge*. According to Tully, it was the failure of this project that pushed Russell to adopt the neutral monist position of Mach and James, which he had previously rejected. Russell's attempt to develop a theory of knowledge, where knowledge is a relation to an external state of affairs, ultimately grounded on the general relation of acquaintance, was a source of personal frustration, until finally, as he relates in a long passage from *My Philosophical Development*, he realized that the relation of acquaintance had to go, although here he puts the emphasis on the superior economy of neutral monism and the lack of introspective evidence for an act of acquaintance:

> During 1918 my view as to mental events underwent a very important change. I had originally accepted Brentano's view that in sensation there are three elements: act, content and object. I had come to think that the distinction of content and object is unnecessary, but I still thought that sensation is a fundamentally relational occurrence in which a subject is "aware" of an object. I had used the concept "awareness" or "acquaintance" to express this relation of subject and object, and had regarded it as fundamental in the theory of empirical knowledge, but I became gradually more doubtful as to this relational character of mental occurrences. In my lectures on logical Atomism I expressed this doubt, but soon after I gave these lectures I became convinced that William James had been right in denying the relational character of sensations ... The contrary view which I came to adopt was first published in 1919 in a paper I read before the Aristotelian Society called "On Propositions: What They Are and How They Mean".
>
> ... In the *Analysis of Mind* I explicitly abandoned "sense data". I said: "Sensations are obviously the source of our knowledge of the world. When, say, I see a person I know coming towards me in the street, it seems as though the mere seeing were knowledge. It is of course undeniable that knowledge comes through the seeing, but I think it is a mistake to regard the mere seeing itself as knowledge. If we are so to regard it, we must distinguish the seeing from what is seen: we must say that, when we see a

patch of color of a certain shape, the patch of color is one thing and our seeing of it is another. This view, however, demands the admission of the subject, or act ... If there is a subject, it can have a relation to the patch of color, namely the sort of relation which we might call awareness. In that case the sensation, as a mental event, will consist of awareness of the color, while the color itself will remain wholly physical and may be called a sense datum to distinguish it from the sensation. The subject, however, appears to be a logical fiction, like mathematical points and instants. It is introduced, not because observation reveals it, but because it is linguistically convenient and apparently demanded by grammar ... If we are to avoid a perfectly gratuitous assumption, we must dispense with the subject as one of the actual ingredients of the world. But when we do this, the possibility of distinguishing the sensation from the sense-datum vanishes; at least I see no way of preserving the distinction. Accordingly the sensation that we have when we see a patch of color simply is that patch of color, an actual constituent of the physical world and part of what physics is concerned with. (Russell 1959/1997, pp. 134–135)

Why was this conversion so difficult, when even in his 1913 *Monist* articles, Russell himself already recognized the sense datum as something physical and had cited the "notable simplification" of the Mach–James view (Russell 1914/1984, p. 21)? He also understood well the difficulties of mental introspection and the challenge Mach and James had raised to it. Clearly these arguments were not new, but they failed to move Russell in 1913 because of other reasons he thought stronger at the time, or because he was still influenced more strongly by Brentano. Tully (1993, p. 20) does *not* accept Russell's whole *Autobiography* story at face value, although he prefaces his paper by noting Russell's explanation and saying he is going beyond the usually accepted reasons for the conversion given above. In my view, the question is not why Russell gave up acquaintance, but why he held to it as long as he did, and what exactly he thought it was accomplishing. Put another way: had Russell's *Theory of Knowledge* manuscript been a success would he ever have given up the theory of acquaintance?

In his 1913 *Monist* articles, Russell also ridiculed the rival causal theory of knowledge he found in Mach and James, illustrated by James's Memorial Hall example, where the Hall gets its "knowing office" by being causally linked to the external object and not from any intentional act of representation. The counterexample Russell gives is of a man who goes looking for his dog in the park, and who on the way falls down a coal cellar where the dog has also fallen (Russell 1914/1984, p. 26). Does that make the man's earlier mental representation of the dog playing in the park instead into a mental representation of a dog in a coal cellar, since that is where the causal chain has successfully led? This objection clearly fails to touch the matter, since Mach and James held that the same causal links that lead to knowledge in one case can indeed lead to *error* in other cases. For example, James says we would "reject a jaguar" if our thoughts connected with tigers causally led us there (James 1977, p. 155) and there is nothing special about success that makes a mental "representation" automatically true of whatever the causal links lead us to, since causal links might just as easily lead us to conclude that the image we originally formed is in error, which is *also* a useful piece of knowledge (Mach 1905/1976, p. 84). Russell's counterexample shows instead how unable he was to take the causal theory seriously, because he was in the grip of *another* theory, namely that Brentanesque theory whereby thoughts or propositions relate "intentionally" or "representatively" to facts in the world. In fact, he even says that when a belief leads to its fulfillment "there must be a logical relation between what is believed in the earlier stages and what is experienced in the fulfillment" (Russell 1914/1984, p. 27). How could this ever be the case in a theory of empirical knowledge?

Another issue to which Russell called explicit attention was that of false beliefs and false complexes of ideas, which he mentioned in *Problems of Philosophy* (1912, ch. 12, p. 92.) According to Tully this concern went back as early as 1907 (Tully 1993, p. 24) and also featured prominently in the *Logical Atomism* lectures (1918/1985), where he credits Wittgenstein with the idea that "You cannot get in space any occurrence which is logically of the same form as a false belief" (Tully 1993, pp. 24–25). Consider, if you take a complex of believed ideas to be real "in itself" then every belief comes out true, at least of itself, and you seem to lend credence to the reality of plainly false beliefs, like ghosts or $2 + 2 = 5$, or to their related complexes of ideas anyway, which do not seem to hold anywhere in nature nor do they obey natural and logical laws (Russell 1914/1984, pp. 24–25).

For Mach and James, this problem of false belief, like so-called sense illusions, is a pseudo-problem, since mental images do not have any representative powers in themselves and are neither true *nor* false of anything as they stand. False belief is a complex of sensations or images which exist in themselves but fail to connect up causally in some expected way, such as a mental fire which fails to burn real sticks. For Mach and James, a belief, or even an hallucination, is not literally what it seems to be. It is not literally a *belief* that there *are* somewhere objects out there external to the experience itself that disobey the laws of physics. The only thing that makes an image about anything, or lends anything as its "knowing office," is a *causal* relation according to Mach and James. A complex of ideas that is "false" just causally leads us to the wrong states of affairs or physical objects, but it leads to other, legitimate causal relations with other mental phenomena like memories or thoughts, and/or the real and physical internal brain energies associated with the hallucination. For example, when Scrooge suggests that Marley's ghost is an undigested piece of potato that could even be true if the extended causal relations led us not to a flickering object in front of Scrooge's eyeballs but to a process of digestion in his stomach and its effects on his brain. As we shall see, even after his conversion, Russell never came around to this "pure" neutral monist view of knowledge and he continued to worry about false, but still representative, complexes of images in "image-propositions."

Here is an alternative story of Russell's neutral monist conversion: problems Russell eventually encountered in formulating this theory of knowledge as belief in his multiple relations theory, strong criticism of the theory by Ludwig Wittgenstein, and self-criticism eventually convinced him to give up on the manuscript, and to adopt a new, and somewhat dualistic, theory of propositions in "On Propositions" of 1919. He then announces that he no longer believes in the theory of acquaintance, but that he *does* now believe in psychologically realistic, but physically unrealistic, "image propositions," and thus still in a kind of propositional theory of knowing, which he salvages in the *Analysis of Mind* (1921, Lectures 8–10). Thus, problems formulating a theory of knowledge may be what really motivated Russell's conversion, as Tully insists (1993, pp. 19–20). Notice the central role the nature of propositions played in the conversion, and *not* primarily the issues of parsimony or introspection. For Russell, the conversion to neutral monism is mostly a conversion to a new theory of propositional knowledge, which he thought could solve the numerous problems of the multiple relations theory, including that of false complexes of belief. The stated reasons for the conversion therefore seem inadequate, as Russell's new theory of propositions in 1919 and 1921 is, if anything, *less* parsimonious because a new entity, the "image proposition," has been introduced, and to boot, the only evidence for this new entity is introspective, which is as unreliable as ever.

The story about all of the problems with *Theory of Knowledge* is rich and I cannot cover all of the details Tully explores in his (1993), or those uncovered by more recent authors such as Landini (2007). One issue clearly has to do with how we formally understand multiple relations, or relations embedded in other relations. As Tully relates (1993, pp. 25–29), Russell still held the following realistic theory in 1914, in some form. If Othello believes that Desdemona loves Cassio, this is a *real* four-place relation Bxyzw where the four slots are filled by the *real* person Othello, the *real* persons Desdemona and Cassio and the *real* relation of loving:

$$B\{O, D, C, l\}$$

The entries in the relation are the actual things and relations and not, for example, the Platonic–Fregean sense, or even psychological mental images. Their behavior is supposed to be factual. Tully says that "Russell's theory of belief was clearly designed from the beginning to anchor a realistic version of the correspondence theory of truth between judgments and facts" (1993, p. 26). Of course, this is just what Russell's theory of descriptions had accomplished for denoting phrases, even in the case of statements about objects that do not exist. He was able to rephrase those statements so that the resulting propositions were perfectly factual, and truth-functional, and did not require the assumption of non-natural Meinongian objects or Fregean senses, and he probably thought that he could continue to deal with belief extensionally if he could just hit on its correct logical form as a relation.

Russell's first idea is that the sentence should be parsed so that Othello believes in a single fact, or complex, namely the unitary fact of Desdemona's-loving-Cassio, or Desdemona's-love-for-Cassio, where the hyphens signify that the belief literally names a single fact made up of the things Desdemona, Cassio, and the relation of loving (Landini 2007, p. 53). This, for the time being, avoids the purely logical, or grammatical, problem of having a relation contain a multi-place relation like loving entered in one of its single slots in the four-place belief relation.

G. F. Stout (1911) then raised an objection, now known as the "narrow direction problem." We want to say Othello believes in the single structured fact of Desdemona's-loving-Cassio. But in order to do so, the relation "loving" is being used in a substantive or "timeless" way when it is simply named as if it were a simple object. In the four-place belief relation, the relation "loving" fills a single slot on the same level as Cassio and Desdemona, as if we were simply naming the relation as a whole object at once. The further complex structure of a two-place relation with its two senses or directions, "x loves y" as opposed to "y loves x," is lost. Tully attributes the criticism to Wittgenstein as well:

> Wittgenstein made Russell face a deeper problem with his theory, one which affects any belief, whether true or false. If the embedded relation is

a mere "brick" on the same level as the two terms it is believed to relate, the model fails to show the specific relation that is supposed to hold between these terms, since it would fail to distinguish between Othello's actual belief and the non-existent belief that Cassio loves Desdemona. Worse, what is to prevent the model from expressing an incoherency like Othello believes that love Cassios Desdemona? (Tully 1993, pp. 26–27)

Russell then advanced a "revised" theory (see especially Russell 1914/1984, pp. 117–118) that included a so-called relational complex, so that the "sense" of the loving relation, l, is uniquely specified:

$$B\{O, D, C, l, R(x,y)\}$$

To accomplish this, Russell adds a technical device for specifying the "what goes where" in the relation. He uses a definite description of the two-place relation (see Landini 2007, p. 59) which involves simpler "place predicates," of a non-complex term and a different, complex term, where the place predication can only be understood in one sense, as predicate to subject, similar to the way the term mortality is predicable of Socrates but not vice versa. So if you encountered Socrates, mortality, and the form of predication in a static fact, presumably you would, supposedly, immediately know what to do with these shaped pieces, as in a static jigsaw puzzle, since only one structure is possible. In a nice reveal, however (Russell 1914/1984, p. 128), Russell also seems to say that the analysis of something so primitive as the order of "xRy" really defies analysis and that "he does not know" how to show how a relation relates its terms. This is just one of the evident tensions in the *Theory of Knowledge*, even as Russell is writing it.

Ludwig Wittgenstein now abruptly enters the story (perhaps in the role of Mephistopheles). He obtained and read Russell's *Theory of Knowledge* manuscript in draft and immediately attacked it. The main objection is rather obscure and Russell didn't understand it at first, but I will quote it from the 1913 letter from Wittgenstein:

> I can now express my objection to your theory of judgment exactly: I believe it is obvious that, from the proposition "A judges that (say) a is in a relation R to b", if correctly analysed, the proposition "aRb ∨ ~aRb" must follow directly without the use of any other premiss. This condition is not fulfilled by your theory. (Wittgenstein 1995)

What is the problem here? According to Landini (2007, pp. 67–68), even if Russell's paraphrase of two-place relations in terms of place predicates were accepted, we still have a sentence like S Believes (Socrates, Mortality,

Predication), where the logical notion of predication of mortality of an individual Socrates is assumed to be immediately understood. But mortality, Socrates, and predication are not really distinguished at all, filling the same slots in the belief relation. They are all still behaving like substantives or nouns, with no verbs to explain what is supposed to be going on. It still isn't clear, because Russell did not make it clear according to Wittgenstein, why "mortality" should *not* be understood as another noun just like Socrates, or why predication should also be understood substantively, without doing the actual *work* of "predicating" one thing of another. Landini claims this distinction is the essence of what Russell learned from Wittgenstein's critique, according to a secondary report by Frank Ramsey (Landini 2007, p. 65). The objection is also similar to Stout's original objection about two-place relations identified "timelessly," but this time affecting the more complex theory involving predication in a definite description of the relation and its places. The criticism can be applied to whatever fills the slots of the belief relation in that so-called "timeless" way. Relations, Predication, Adjectives, Verbs, and Properties, when entered as substantives in a timeless fact, do not do their actual work of relating, predicating, modifying, doing, and so forth. This criticism may have been crucial to Russell's change of mind about the viability of the theory of multiple relations, also bringing about a strong emotional reaction, about which more below.

Even if these difficulties could be met by some other theory, as Russell says, it still leaves the problem of how these relational complexes are known or believed (Russell 1914/1984, p. 128). Is the relation of loving the same when embedded in the context of being believed, as it is when it holds between people or objects outside the context of belief? I cannot see how Russell could ever have answered this objection, since it seems clear the relation is simply *not* the same one inside and outside, and this is a problem that is well known to analytic philosophers today. When you try to embed a relation like "loving" in another relation like "belief" you actually don't embed it at all, but change it into *another* relation of "believed loving" on a logically independent footing, blocking logical inferences about the real relation holding between the real persons. And of course even that is probably wrong, because Othello's belief is supposed to be about Cassio's *loving* Desdemona, not his own mental state of *believed* loving. Why would Othello be upset about his own imagined complexes of ideas being a certain way?

This latter line of questioning, which may also have received some impetus from Wittgenstein (see Landini 2007, p. 69) gives us, so I think at any rate, an important clue about Russell's change of mind. By the time

of "On Propositions" (1919) Russell realized he could save his wounded
realist program, but only by distinguishing the real objects from the
believed objects, calling the latter "psychological phenomena" or images,
and treating their relations according to psychological laws, not physical
laws, nor presumably extensional laws of logic either. The believed proposition is now what Russell called an "image proposition" consisting
of mental images subject *only* to the laws of psychology, even if he says
in a footnote that this division between psychology and physics may only
be provisional (Russell 1919, p. 18n.). The idea is to treat believed image
propositions in intentional contexts as truths about the workings of human
psychology, rather than truths about the physical world. Believed relations
and real relations lie in separate, parallel orders and obey different laws
(which Russell says explicitly here and in the 1921 *Analysis of Mind*, Lecture
8). He bit the bullet, therefore, and proposed that these are just two
different relations "loving" holding among mind-external things like
Cassio and Desdemona, and "believed loving" among Othello's mental
images, so that in effect one is not literally embedded in the other.

For all that, Russell took the belief context as *psychologically* realistic.
This is perhaps the real reason why he found neutral monism attractive
again in 1919, with its two mental and physical orders of variation for the
same neutral contents, or so I think anyway. Neutral monism seemed to
Russell to solve the problem of believed propositions and to save his theory
of knowledge. But this desperate move also destroyed any equivalence at all
between the believed "image proposition" holding among the mental
images and a proposition about real physical objects and their extensional
relations, so how could a believed proposition made up of mental images
ever constitute *knowledge* of an external state of affairs?

Russell was desperate because by 1916 the *Theory of Knowledge*, and
the theory of acquaintance at its foundation, was clearly a degenerating
research program and Russell was in despair about it. This was the
situation of which he wrote in the aforementioned 1916 letter to Ottoline
Morrell:

> I wrote a lot of stuff about Theory of Knowledge, which Wittgenstein
> criticized with the greatest severity ... His criticism was an event of first-
> rate importance in my life and affected everything I have done since. I saw
> he was right, and I saw that I could not hope ever again to do fundamental
> work in philosophy (1968/1998, p. 57)

This is the quote which has been seized upon as "evidence" that Russell
was through with philosophy and had all but turned the subject over to

Wittgenstein. But this is just a snapshot in time, only two years or so after the *Theory of Knowledge* manuscript and before Russell's works the *Analysis of Mind* and the *Analysis of Matter*, when his philosophical powers had clearly returned (as Bostock 2012 points out).

Recall that Russell, as a realist, could not accept Frege's Platonic theory of abstract propositions, calling them "unrealities" and "fictions" (Tully 1993, p. 200). The Fregean theory required a third thing between the belief, the proposition believed, and the values True or False, or at least those states of affairs or arrangements of things making the proposition true or false. Wittgenstein's *Tractatus* would give free rein to this impulse, invoking the device of a Platonic "logical space" to explain both the sense of propositions in linguistic expressions, or even combinatorial relations of symbols or letters, and arrangements of factual states of affairs in the same abstract "space" of combinations. Wittgenstein solves the problem of how to understand propositions in a neutral way but at the high cost of adopting a logical Platonism, a dogma of picturing or representation of one complex by another, and abandoning realism about psychology and human knowledge on which Russell had insisted, making propositions into almost inhumanly perfect entities there is little reason to think have anything do with our actual thoughts and beliefs. Who thinks our beliefs are really about abstract combinations of letters, or concepts, in logical space? I suggest that faced with this impasse—accept abstract propositions or psychologize propositions to make them real in some natural domain—Russell chose the latter. The "image proposition" was clearly not one of Russell's better ideas, but it was better, for him, than the unreal alternatives offered by Frege or Wittgenstein.

Refuge in psychologism: a not-quite-neutral monism?

So, in "On Propositions" (1919) and in the *Analysis of Mind* (1921), Russell returned to the drawing board and reframed his theory of knowledge. As Russell says, the neutral particulars, whether particular sensations or mental images even, are now *neither* mental *nor* physical, but more fundamental than either:

> The stuff of which our world is composed is, in my belief, neither mind nor matter, but something more primitive than either. Both mind and matter seem to be composite, and the stuff of which they are compounded lies in a sense between the two, in a sense above them like a common ancestor. (Russell 1921, pp. 10–11)

The particulars are neutral but they can be arranged in different orders and in obedience to different laws. One order is of physical laws and causal relations and the other order is of psychological laws and associations, but the same neutral particular may participate in both. In themselves, even mental images do not differ in any intrinsic way from other neutral events, except that they are related in different complexes by psychological variations such as trains of association, memories, thoughts, and the like, which physical events following physical laws do not obey. For Russell, however, when we divvy up the common elements into special classes, we always end up putting mental images in the mental category and physical items in the physical category: "images belong only to the mental world, while those occurrences (if any) which do not form part of any experience belong only to the physical world" (Russell 1921, p. 25). Mach and James would never have said this. This new view has a number of very odd features. Entities in complexes of "image propositions" are "purely mental" (Russell 1919, p. 18). The complexes obey only the peculiar psycho-logic of belief and its non-extensional rules of association. He *also* says, however, that he does not know whether the distinction will be ultimate and irreducible (p. 18n).

Russell still clearly thinks image propositions have that "intentional" power to represent or picture mind-external facts, despite claiming to have freed himself from Brentano:

> Images, in accordance with what has just been said, are not to be regarded as relational in their own nature; nevertheless, at least in the case of memory images, they are felt to point beyond themselves to something which they "mean." (Russell 1919, p. 26)

But his ongoing fear of the neutral monism of Mach and James still seems to be: if we regard psychological phenomena as real in "themselves," then the complexes they seem to represent will be real. So any false complex of images will be taken as true of something, which they are not. On the other hand, image propositions cannot lose *all* of their representational powers for Russell, for they have to be capable of representing external states of affairs correctly if believing in these image propositions could ever constitute knowledge. They cannot be said to "represent nothing truly or falsely" à la Mach, or James, for that would remove the power of a believed image proposition to be true or false, in those cases where knowledge is actually attained through mediate belief in a proposition. This means Russell must take the denizens of image propositions in a representative sense, but if he does so, then he simply cannot take the interrelations of images in a proposition to be genuine realities in themselves, as Mach and James do; they have to represent some parallel order of images that cannot be the same as the physical order, else the false situations they represent would have to be true, representatively, of something. A revealing passage from the 1918 *Philosophy of Logical Atomism* is the following:

> You will notice that whenever one gets to really close quarters with the theory of error one has the puzzle of how to deal with error without assuming the existence of the non-existent. I mean that every theory of error sooner or later wrecks itself by assuming the existence of the non-existent. (Russell 1918/1985, p. 90)

True perhaps for every theory that *also* assumes a representational theory of the image or image proposition. Russell never even considers getting rid of representation for images and image propositions. Given that Russell believed the distinction between physics and psychology was provisional only, however, I wonder if he ever considered that these sorts of strange psychological complexes of images *could* be realized directly in a physical sense in the brain, as Mach says of hallucinations, so long as we are clear that these real phenomena represent *nothing* truly or falsely outside themselves and that even their strangest behavior is perfectly comprehensible as the physical behavior of internal brain energies. Recall also James's remark above that mental "work" and "action" are just as real and energetic in their own way in describing physiological brain processes as physics is true of events elsewhere in nature (James 1977, p. 289). James and Mach do already have an error theory that is not vulnerable to Russell's problem, but he doesn't see it; their causal theory of knowledge *and* error is simply closed to him. Thus the weird psychophysical "dualism" of the *Analysis of Mind* is the only realistic way out left to Russell and he embraces it.

In sum, this quandary of issues, I suggest, shows Russell's dyed-in-the-wool commitment to representational or propositional theories of knowing, even despite the rejection of the *Theory of Knowledge*. In his lecture on "Belief" in the *Analysis of Mind* (Lecture 12) Russell is still using the "image propositions" to convey true or false beliefs that constitute knowledge. Nothing has really changed except that now the psychologically realistic "image proposition" has become the intermediary representative between belief and the facts that make beliefs true or false. The philosophical problems with this view seem as severe as ever. Propositions as objects of belief have to be psychologically realistic, and obey the peculiar psychologic of belief, but they also have to bridge to the extensional features of the world when the beliefs are true of something else, and there seems no way the same proposition could make the same statements and do both things at once. Better at this point to simply get rid of propositions altogether and make the move to a causal theory: the complexes of images, stripped of representative powers, could then be tied to *both* kinds of variation simultaneously, to their own psychological variations as brain energies taken in themselves and to external objects via causal links.

Mentalistic, but at least "psychologically realistic," image propositions thus resided in an unhappy halfway house between Russell's old realistic theory and the more thorough neutral monism and causal theory of knowledge and error of Mach and James. The problem, as we predicted, is that a bifurcation has once again opened up between mental phenomena, their laws and true propositions about them, and physical phenomena and propositions true of that world. In effect, Russell had become a dualist and not a neutral monist after all (as Tully also holds: 1993, p. 34), this time a "propositional dualist" about believed propositions and factual states of affairs, or the "two orders" of physical and psychological realms. Even if he says the distinction might eventually be overcome he takes no steps in this direction.

There was thus a rather great difference between what Russell was calling "neutral monism" and the position of Mach and James on the theory of knowledge. This is an objection to my thesis that Russell's neutral monism should be treated as of a piece with the realistic empiricism of Mach and James (as I also noted in my 2003, p. 153). Russell's representative assumption about the "intrinsic intentionality" of mental images is just alien to Mach and James. And if we have to choose sides, it certainly seems that the causal theory is far more naturalistic and true to life and experience and thus preferable to representative propositional theories of knowledge. The strange thing is that Russell's naturalism did not lead him down this road, even as he became a model naturalist in other ways.

Russell's system of perspectives and external world program

Russell's neutral monism was also developing out of his philosophy of science in a much more straightforward way. In *Our Knowledge of the External World* (1913/1926) Russell built up what he called a system of "perspectives": a construction of physical space and objects out of sensibilia (sense data if we are acquainted with them), that is, events, bound in causal-perspectival relations with each other. He says that the object in space is really the sum of its aspects, or "what common sense would call its 'effects' upon, or interactions with, observers," as he says quite clearly later in the 1921 *Analysis of Mind* in his Sirius example (1921, pp. 101–102). At different points in space, an object like a penny will cast different images, some circular, elliptical, wedge-like, and so forth, but in all of these aspects, the individual events making up what we call the "interactions" of the penny with the environment will be related by solid laws of perspective (1913/1926, p. 74). So, for example, when the observer approaches the penny along the same angle of sight the penny gets larger but its shape remains the same (this is a law grouping those events together into the penny's image). Or the size is constant with the distance from the penny but the shape changes with different angles of sight. The perspective represents the observer's causal interactions with the object, and so each perspective is not just an optical perspective, although it is that too, but a complete catalogue of causal effects, similar to Mach's general "causal–functional" maps of his elements.

As Russell puts it, each spatial perspective is a three-dimensional snapshot we can order in a higher six-dimensional space, each *point* of which is a three-dimensional perspective. What happens when we artificially divide physics from psychology is that we take the events marked physical and construct a space consisting only of those events and perspectives, including even all events in brains that are externally measured or observed by devices, but *excluding* the actual sensations or other events as the possessor of the brain experiences them directly. Those experienced events are then grouped into a different space of perspectives called private, psychological spaces of touch, sound, sight, etc. In reality these spaces are not separated at all. Every event finds a place in the general six-dimensional theory of perspectives, just as in Mach's original causal–functional maps. As Russell insists, the spaces of physics (all natural events minus sensations or other mental events) and the space of psychology (minus the mind-external physical events) are sections taken out of the more fundamental general space of events and causal perspectives. The neutral event particulars are neither mental nor physical; they become "mental" when arranged into psychological spaces and fields of perception and they become "physical" when arranged in a four-dimensional space-time of physical space and objects.

Physical objects are decomposed into interaction events entered into a consistent set of perspectives. The events consisting of the commonsensical "effects" of the penny are gathered into the perspectives on the penny. As we get closer to the penny its effects become more intense, but we never arrive at a point where we have to do with a penny *an sich*, the substance or nucleus absent all external effects. In fact, there is no penny *an sich*, all we would find there upon interacting with it more closely are more effects, especially intense atomic "effects," commonsensically speaking, of the pieces of matter making up the penny.

This feature of Russell's view baffled critics from Arthur Lovejoy to W. T. Stace, who were still inclined to see the *actual penny* as an object or nucleus situated in the six-dimensional space of perspectives. Russell may have added to the confusion (1927/1954, p. 211) when he said, tongue in cheek, that objects are "empty-centered," there being nothing at the center of the perspective space of a group of percepts: "A group is hollow: when we get sufficiently near to its centre it ceases to have members." But it is not "hollow," at each point there is always a perspective of some kind. Only objects in conventionally separated physics and psychology will be hollow and have missing perspectives. The physical brain will lack sensations, having left them out of consideration, and the psychological perspectives will lack physical objects and consist only of bundles of sensations. Russell should rather have said, in my opinion anyway, that there are no such things as "centers" in the broader perspective space, because there are no such things as the nuclei or centers of objects at all, except as an ordering of the individual events of interaction into perspectives. Thus there is no "empty" missing perspective at the center of the object, or perspectives outside the space looking in; all viewpoints are internal ones. The space of perspectives is continuous with a perspective at every point, and has no internal holes or external boundaries.

In a second edition of *Our Knowledge of the External World* in 1926, Russell explicitly referred to the Heisenberg matrix atom as exemplifying his object of physical inquiry as an event-based construction, a matrix of its constituent events, rather than a thing-based object or nucleus. Perhaps this was why he favored the empty-center view of objects, since he thought atoms were already being conceived that way in physics:

> It was always assumed that there is something indestructible which is capable of motion in space; what is indestructible was always very small but did not always occupy a mere point in space. This view still dominated the Rutherford–Bohr theory of the structure of the atom. Since 1925, however, under the influence of de Broglie, Heisenberg and Schrödinger, physicists have been led to resolve the atom into systems of wave motions, or radiations coming from the place where the atom was supposed to be. This change has brought physics closer to psychology, since the supposed material units are now merely logical constructions. In regard to space and time, relativity has introduced a fundamental structural change by merging

them in the one four dimensional space-time. Both these changes have
made physics easier to reconcile with psychology than was formerly the
case. Both sciences now demand certain departures from common-sense
metaphysics, and fortunately the departures they demand harmonize with
each other. (1913/1926, p. 83)

Russell was deeply influenced by the theory of relativity in his construction
of the theory of perspectives from the individual aspects or concrete events
of observation ordered by light cones emanating from every point in space.
The causal theory of space by A. A. Robb (1913), for example, seems to
have made a deep impression on him. For Russell, objects and the
biographies of objects are just worldlines consisting of events, making up
the interactions of the object with all others along causal lines or light rays,
again eliminating the object or substance, with its intrinsic properties, as
unscientific and unverifiable in principle. In the case of special relativity
this involved a causal theory of time and events, where the metrical
structure of space is determined by exchanging light signals, where every
such exchange, emission, absorption, reflection is an event, or at least a
potential event. In the case of the atom, the concrete transitions between
potential energy levels are not merely the only *observable* quantities, as
Heisenberg had insisted in his 1925 paper; Russell implies here that they
may also be the only concretely *existing* entities making up the atom itself
as a series of causal perspectives. The solidity or indestructibility we seem
to see there is rather the solidity of a law or function but not of a classical
billiard ball atom or nucleus.

The *Analysis of Matter*: event particulars and the controversy with Newman

Russell considered the 1927 *Analysis of Matter* a companion volume to the
1921 *Analysis of Mind*. The book is unforgiving in its mathematical and
scientific detail, showing Russell's strong grasp of the material. There
are brief treatments of tensor analysis in general relativity and expositions
of Heisenberg, Born, and Jordan's matrix mechanics. While many commentators considered the *Analysis of Matter* to mark a shift to a new
"scientific realist" stage in Russell's thinking, he denied this, and, as Tully
(1993) and Lockwood (1981) point out, the change of terminology, "sense
datum" to "percept," "sensibile" to "event particular," "effect" or "interaction" to "quality" or "intrinsic character of matter," really does
not affect the neutral stuff under discussion, only the changes that were
due to the acquaintance relation, which has finally been dropped. As Tully

pointedly complains, the physicalistic sounding word "event particular" is *not* exclusively physical, excluding psychology, but neutral for Russell, and thus would also cover mental events or "percepts" as event particulars which occur in human brains. The words "event" and "percept" simply indicate provisional distinctions between inquiries in physics or psychology, not fundamentally different categories of things. The goal therefore is hardly to "abandon" neutral monism, but to enhance and expand its vistas, applying the view broadly as the logical skeleton of a scientific philosophy or general metaphysics of event particulars. In other words, Russell probably saw the view as an umbrella view or theory schema, as Mach had, and attempted to use it to unite aspects of early twentieth-century science: the theory of general relativity and quantum mechanics in particular, with the equally objective facts of human perception. A general unification of science is in view.

Russell saw two main applications of neutral monism where the view could be especially useful. First, it must explain the relation of sensations to the physical world in which sensations naturally occur as physical phenomena *without* contradicting the results of physics. Second, it must articulate a scientific realism based upon the event particular and perspectival relations and obtaining objects like extended atoms, bodies, and physical space and time from events and relations of some kind.

As part of his neutral monist enhancement of physics, he then makes a very famous observation that physics describes the mathematical causal skeleton of the world or a network of abstract relations of pieces of matter in space-time, but says nothing about the "intrinsic character" of those events. Indeed, Russell admits, we have no reason to think it *unlike* the concrete qualities manifested in our percepts, our sensations, except that the particular qualities of matter, or rather the sum of its effects, are unknown to us.

> Physics itself is exceedingly abstract and reveals only mathematical characteristics of the material with which it deals. It does not tell us anything as to the intrinsic character of this material. Psychology is preferable in this respect but it is not causally autonomous ... But by bringing physics and psychology together, we are able to include psychical events in the material of physics and to give physics the greater concreteness which results from our more intimate acquaintance with the subject matter of our own experience. (1927/1954, p. 10)

> As to intrinsic character, we do not know enough about it in the physical world to have a right to say that it is very different from that of percepts; while as to structure we have reason to hold that it is similar in the stimulus and the percept. (1927/1954, pp. 400–401; a similar passage occurs in Kant's paralogisms A 358–9)

In essence, Russell is suggesting that for a full account of the physical world that includes psychology, we need to take note of what concrete material the physical properties such as mass or charge are instantiated in. The properties of physics permit this because they are abstract and apply generally, but physics remains silent as to what instantiates them. We are, however, capable of knowing the intrinsic character of those events that go on in our own brains: our sensations, with their qualities, are physical events in the brain whose nature is expressed directly through their qualities and not simply through their structural arrangement (1927/ 1954, p. 320).

In a later exchange, the mathematician Max Newman criticized Russell for his purely structural view of conventional physics. Newman pointed out that physics was not, and could not be, entirely a set of abstract relations of things because this description is far too thin to pin down any concrete instantiation for laws, let alone objects and experiments. *Any set of things can be arranged in any number of abstract relations or combinations whatever, limited only by the cardinal number of objects one starts with.* How would one specify a specific set unless the relations themselves were concretely instantiated in the qualities of the events? This observation, which Russell accepted immediately, contradicts his all too thin gloss on physics considered as an abstract system. But, as I have urged (Banks 2010), Newman actually makes Russell's point for him: physics *needs* grounding in the concrete "qualities" (Newman's word) of the particular material studied. The events that are related structurally by the system of perspectives into objects in space-time must themselves be concretely instantiated by qualities to pin down the multiplicity of possible abstract relations. The conclusion is that an "intrinsic character" of events is not so much an add-on, or a useless and isolated metaphysical hypothesis, as it is often taken to be (see Grayling 2003), but rather a necessary addition to any purely mathematical theory of physical events, effectively grounding their structural relations. Thus Russell was not merely a structural realist about physics, as many would read him, he was also a realist about whatever instantiated those structural relations, and this was certainly true after the exchange with Newman. It all depends on whether "physics" means the physics of matter, space, and time, but excluding sensations or any other qualitative grounding, "physics" in the conventional sense, or whether "physics" was being used in the enhanced sense Russell sometimes adopted, in which sensations *and* the presently unknown "intrinsic character" or qualitative grounding for relations of matter would be included.

Russell's constructivist project: space-time and the external world

After the delineation of the neutral stuff as event particulars with intrinsic concrete quality and dynamical effect, Russell began a constructive project, employing his considerable powers of deduction. He used event particulars to carry out mathematical constructions of the manifold of space-time, and the construction of matter in quantum theory, to make neutral monism compatible with physics. This he saw as a job for mathematical logic par excellence, which was compatible with his realistic bent of thought and his neutral monist project. Certainly the naturalistic turn sharply separates Russell from Wittgenstein, whose linguistic anti-naturalism only became more and more pronounced. The history of analytic philosophy is understandably more focused on this whole movement toward the linguistic turn, the second-order analysis of language and logic of scientific discourse of the sort that went on in the Vienna Circle, but I feel this will prove to be much less interesting in retrospect when compared with the slow but inexorable movement toward a truly naturalistic philosophy in the twentieth century, of which Russell is the clear forerunner (see Quine 1966).

Russell's constructive project had begun already in *Our Knowledge of the External World*, in which he used overlapping events or qualities such as color patches to define points, a procedure that he attributed to his *Principia Mathematica* collaborator, Alfred North Whitehead. "Points" are defined as limits of overlapping events, similar to what is done in topology, where concepts like the interior points of a region and boundary points are defined by overlapping neighborhoods and their set-theoretical relations. What is supposed to favor Russell's and Whitehead's construction is that the structure of the topological manifold of space-time in relativity results from causal relations between real physical events in the manifold. We do not therefore need to assume a prior space of abstract mathematical points and regions and then embed physical events in it.

There are several problems with this, however. A closer look at the construction reveals that the events must already be of *some* space and time extension in order to fill their role of topological neighborhoods, although it is true that these may be taken as small as we wish. It is also true that the structure of space-time, the metric, is not yet defined on this extension, but the raw extension must be already there for the construction to work. Russell's construction is thus not an explicit definition or analysis of the property of "being extended" in space or time. The notions of "overlap" and all of the constructions are already processes carried out in pre-existing space-time. Rather, Russell's "construction of space" belongs to a later stage of identifying points in space and laying down a continuous manifold of connected points in which the structure of neighborhoods is always preserved no matter how this manifold is stretched or contorted, say by the presence of matter in the theory of relativity. But the *origin* of extension is as mysterious as ever, as is the construction of extended space, time, and matter from unextended event particulars, which simply is not pursued by Russell or by any of the other constructivists who developed the causal theory of time after him (such as Reichenbach 1927/1956, Winnie 1977). For example, as many critics have observed, both Reichenbach's "mark" theory and Winnie's "primitive asymmetric causal relation" are dependent on a prior *extended* view of causal relations and processes already in space or time, and thus cannot be used to analyze these notions.

Russell is forced into a corner when he has to call on the dimensionality of this background space to constrain the possible overlaps by fiat. This occurs in his definition of a "quintet" of events in the *Analysis of Matter*. Five overlapping space-time events *must* share a point because the background manifold of space-time is four-dimensional, a blob of points in R^4. Moreover, overlaps defining the relation of "between-ness" have to be stipulated such that higher dimensional "re-entrant" events are explicitly ruled out, also by stipulation (1927/1954, p. 305). These assumptions are stated in the form of axioms, but I think the reason he has to state them as axioms is because they involve appeals to extended intuition which Russell is unable to eliminate or analyze into simpler ideas. The dimensionality of the background manifold as well as its extension must be assumed implicitly and intuitively, and what we have is a specific theory of space-time within a more primitive unstructured extension, or perhaps a catalogue of more specific kinds of spaces within an overarching spatial extension, but *not* a theory of the origin of basic concepts of extension or dimensionality, duration, etc. Hence we must assume that Russell's construction is ultimately unsuccessful, although it points the way to the solution for a problem that realistic empiricist views must face: the construction of space, time, and objects out of instantaneous, "transient" events and causal perspectives.

Russell speaks quite a bit about the natural qualities of events in the *Analysis of Matter*, probably making many readers, like Quine, think he was a phenomenalist or a Humean empiricist operating with sensations. I think, however, that Russellian qualities must be understood in a physicalistic way. Russellian qualities are simply physical "effects" of causal powers manifested in events. Russell clearly thinks event particulars are the strewn "effects" or "interactions" of what common sense calls an object. Effects gathered in certain places in perspective space are events, or events are the sum of these effects registered at a given point in perspective space. Second, Russell thinks events are "quintets" of compresent "qualities" gathered in four-dimensional space-time points or small regions where they overlap. So what he calls the "effects" or "interactions" gathered into events in one context he actually calls "qualities" gathered into events in another context. (I myself think the notion of "interaction" in the metaphysics is being modeled by the topological idea of "overlap" in his logical analysis of events.) Qualities, despite the subjective-sounding overtones, are really just the manifested effects of what common sense would call the causal powers of objects, expressed through interactions with an observer or purely physical interactions with each other. These causal powers are manifested by and in events as the event's quality or "intrinsic character." There is absolutely nothing mysterious, unphysical or "panpsychist" about Russellian "qualities" in matter, they are just the concrete expressions of natural powers, like Mach's elements. Physical qualities occur everywhere in all natural events, but we only experience them in the case of percepts, which are physical events in our brains. This experience is enough to venture a guess by Russell that events elsewhere also manifest qualities or intrinsic character but we have as yet no access to verify what those qualities might be like.

Conclusion: Russellian event particulars

To sum up, then, I believe Russell's neutral event particulars can be characterized as follows:

1. Event particulars are "transient" (1921, pp. 143–144), non-repeating, individual "happenings" that come and go in time, not universal properties or types, and not objects or structures. Objects' worldlines or "biographies" are composed of event particulars in individualized perspectives.
2. Event particulars are causally linked to one another by physical laws, and are themselves what common sense regards as "effects" or manifestations of "interactions" between objects (1921, pp. 101–102). The event particular manifests its concrete qualities through its causal interactions with human observers or other objects. Where common

sense would see relational effects of an object, or atom, or star, like Sirius, the neutral monist sees these so-called effects or interaction events as primary, and uses them to construct the object formally out of event particulars bound together by perspectives, or laws of perspective.

3. Physics in the narrow sense consists of a relational structure grounded in the real concrete qualities, intrinsic character, possessed by event particulars in nature (1927/1954, p. 10) and their causal relations. The quantitative mathematical relationships of physics are concretely instantiated causal relations, not abstract structural relations. The relational structure of physics is ultimately grounded in the qualities or intrinsic character manifested by physical event particulars in what we might call the enhanced physical world view.

4. Event particulars manifest individual qualities which serve to identify them, although we do not know what sort they are except in the case of our own percepts, which are physical events taking place in our brains. However, because physical events in our brains exhibit individual concrete qualities, and they are real denizens of the universe, then so presumably do other physical events and their qualities elsewhere, although we *cannot* infer that the qualities must be similar or related by a line of descent as in panpsychism (1927/1954, pp. 400–401).

If we are willing to overlook Russell's strange continued adherence to the representational-propositional theory of knowledge, I think evidence overwhelmingly warrants collecting the views of this triumvirate of Mach, James, and Russell together and calling them "classic" neutral monism or just realistic empiricism. It is certainly true that Russell's "neutral monist" view of the physical (although dismissed repeatedly, for example by Quine 1966) has been the most influential, and survived into the twentieth century, in versions like Herbert Feigl's 1958 essay *The "Mental" and the "Physical"* (1958, Section 5E in particular). In the 1970s, Russell's theory of the "intrinsic character of matter" was revived by Grover Maxwell (1978) to counter Kripke's modal argument for dualism in *Naming and Necessity*. Michael Lockwood reinterpreted, and updated, Russellian neutral monism in his own path-breaking philosophical writings such as *Mind, Brain and the Quantum* (1989). Finally, David Chalmers suggested taking Russellian monism seriously as an option for supplanting orthodox physicalism in his 1996 *Conscious Mind* and also in his recent collection *The Character of Consciousness* (2010), where he presents neutral monism as an explicit alternative to his own dualistic and epiphenomenalist views, without directly endorsing it. "Neo-Russellian" monist views have made a modest comeback in the philosophy of mind, in a multitude of different forms, some of them panpsychist and others not (Stoljar 2001, Rosenberg 2004, Strawson 2006, Banks 2003, 2010, and many others). I will consider the contemporary versions of this view, as well as my own contribution to the discussion of physicalism, in the next chapter.

CHAPTER 5

Enhanced physicalism

Introduction

In this chapter I will expand upon Russell's hypothesis in the *Analysis of Matter* (1927) that mind-independent natural events have a qualitative intrinsic character, and show how this hypothesis can enhance our view of the natural world by making room in it for the reality of human sensation. I will suggest adopting what I call "enhanced physicalism" (Banks 2010), which seems to me the best extension of Russell's hypothesis, and which has not been represented in the literature on neo-Russellian monism. The main difference between my reconstructed view—which I do not attribute to the historical Russell—and others in the literature will be that I think enhanced physicalism should be an a posteriori physicalism. This means that while sensations and individualized neural events in the brain are the same kind of neutral events, one macro- and one micro-manifestations of the same underlying neural energy, these neural energies are *manifested* differently in the two cases, making any a priori deduction of one empirical manifestation from the other impossible. This chapter will take a very different approach from the way neo-Russellian monism is handled in the philosophy of mind: most of those approaches are panpsychist and implicitly rely on an a priori physicalist species of explanation. The approach taken here starts from the philosophy of science, in particular with the subject of causation, powers, their manifestations in events, macro and micro, and a basic explanatory, but not metaphysical, dualism between two different empirical manifestations of the same powers, based on the example of the conservation of energy. Finally, I will compare my neo-Russellian reconstruction to those of David Chalmers (2002, 2010) and Galen Strawson (2006).

Russell's hypothesis in the *Analysis of Matter*

The problem for any neutral monist view is how to define the neutral event particulars, rather than to say what they are not (as Stubenberg 2010 points out). Recall that Russell claimed event particulars are like the disembodied effects resulting from causal interactions between objects and their environments in the commonsensical way of representing nature. He then turned the tables on common sense and made these interaction events more fundamental than the objects themselves. An event particular is the pure manifestation of an effect, which we sum into a system of interacting objects. We order events into objects, by following the laws of causal perspectives, from the different vantage points in which we take in their effects. There is thus excellent reason to think that Russell's "event particulars" can be understood as manifestations of the effects of powers, manifested in the event's qualities, just as they would be if we considered the effects of an object on its environment of other objects and isolated just those effects to characterize what we mean by an "object." We need only recognize that the events come first before there are objects or fields or egos, despite what common sense, or natural language, places first.

Needless to say, this event-centric view is very different from that of most physicists, who take objects of some sort, particles and fields, to be fundamental, then think of powers as possessed by those objects, and finally events of interaction between one object and another take place at the third level. For Russellian neutral monism, the disembodied effects in which causal powers are manifested are fundamental and these event particulars actually comprise the particles and fields of natural science as well as mental phenomena like sensations (see Banks 2003, ch. 9). The ontology of events is preferred in neutral monism because it has the best chance of uniting sensations, which are clearly brain events of some sort, and events in physics under the same roof. Our attention will therefore be confined to powers and their qualitative manifestations in individual events.

Russell claimed, in the *Analysis of Matter* (1927/1954, p. 10; pp. 400–401), that physics, traditionally conceived, studies the quantitative, mathematical structural relations between events in space and time, but says nothing about the "intrinsic character" of the matter (or rather events making up the matter) in which those relations might be instantiated. In the *special* case of the physical events in our brains called "percepts," however, Russell says, we *do* have direct access to the intrinsic character of those brain events through our sensations. Russell's hypothesis about physics being purely structural was immediately challenged by the mathematician Max Newman (Newman 1928; see also Demopoulos and Friedman 1989 and Seager 2006 for more details of the story). Newman pointed out that physics could not possibly be described as purely relational and lacking any concrete instantiation, without having some kind of "qualitative" grounding (his word). Newman pointed out that given the right cardinal number of items, they could be arranged into any combinatorial way whatever. Therefore, the purely relational content of physics could be realized combinatorially by any items having the right cardinality, ping-pong balls, dominoes, whatever. This removes all experimental content, or grounding, from physics and lifts it off the ground into a purely formal structure, which seems wrong. Indeed, this is more like the logical positivist ideal of physics as an interpreted formal system.

In a letter to Newman (reprinted in Seager 2006), Russell accepted this criticism wholeheartedly and retracted his definition of physics narrowly conceived. It is thus clear from the Russell–Newman exchange that physical laws, properties, and relations certainly have *some* kind of "qualitative" grounding and are not just part of an abstract mathematical or relational structure. But this directs attention to what that qualitative instantiation is and *how* it grounds the causal relations of physical science, which are the main questions we will consider below. As I have pointed out before (Banks 2010) I don't think it was Newman's point to *refute* Russell's hypothesis about the nature of the physical, and the need for a concrete grounding in natural qualities, but rather to give it a sharper content. As we saw above, Russell also used the term "physics" to indicate a future enhanced physics in which relations *would* be instantiated by event particulars with qualities, and sensations would be included in physics along with traditional physical objects and events. It is this idea of an enhanced physics to which I now turn my attention and grant Newman's point that physics could never have been purely structural in nature.

Standard physicalism versus enhanced physicalism

Today, we can express these issues with more precision by adopting some contemporary language about the meaning of physicalism and, I suggest, by adopting the language of powers, their manifestation conditions, and particular events in which power manifestations occur, concepts which I think are well suited for reconstructing the realistic empiricist view along some of the lines suggested by Mach, James, and Russell, although certainly some changes must be made.

Physicalism is the idea that the description of the world given by physics (of particles with mass, spin, and charge moving in fields of force) describes the world *completely*. Everything else (biology, human experience, economic behavior) supervenes on the physical description. This kind of supervenience thesis is taken to be definitive of physicalism (for discussion see Gillett and Loewer 2001, Kim 2005, Stoljar 2010) rather than any definition of basic physical entities or laws. In addition, physicalist supervenience is stated in terms of properties and/or relations. For our purposes, standard physicalism can be defined (following Kim 2005) as a strengthened supervenience thesis in terms of physical (or micro) properties Px and other (macro) properties Mx:

(1) A situation S is standard physicalist if for any system s in S that instantiates Mx at t, (1) there exists a property Px, such that s instantiates Px at t, and (2) necessarily anything instantiating Px at that time instantiates Mx at that time. (Kim 2005, p. 33)

The Px predicates stand for properties like mass, charge, and spin, along with some description of their individual spatio-temporal relations, or configurations. The Mx predicates are higher-order properties or relations, for example mental properties. As Kim points out, the first clause only guarantees that M properties strictly covary with P properties, which even a Cartesian dualist would accept. The second clause is necessary to ensure that the presence of properties Mx necessarily depends upon there being a supervenience base of Px properties, although there may be more than one for any given higher-order property Mx. Then, if you instantiate any supervenience base of Px, you also instantiate Mx.

Notice that standard physicalism is couched at the level of properties and relations and is silent on the question of what exactly instantiates them. Two situations are standard-physically identical if they are identical in all properties and relations and if they are instantiated by some individuals, provided they are the same in Px and Mx. But put in "b" for "a" and you still get a perfectly acceptable supervenience relation, even though you have changed what instantiates it. Thus standard physicalism does not completely pin down the instantiation. I believe this point was first made by Grover Maxwell (1978), namely that a truly physical description should lock in all the way down to the individuals that instantiate physical properties and relations. For this reason, we can introduce an "enhanced physicalism" to fill the gap as follows:

(2) Two situations S_1 and S_2 are identical in the enhanced physical sense if in both S_1 and S_2, for any system s that instantiates Mx at t, (1) there exists a property Px, such that s instantiates Px at t, and (2) necessarily anything instantiating Px at that time instantiates Mx at that time, and (3) necessarily whatever individuals s (events or objects) instantiate Px and Mx in S_1 also do so in S_2.

There will be situations where what instantiates our physical properties and relations is something else, b-matter as opposed to a-matter, and standard physicalism will still hold them identical, but these will not count as identical situations in the *enhanced* physical sense.

Even in standard physicalism some particular content was always meant to instantiate physical properties, as Newman rightly insisted. For example, in nineteenth-century mechanism, the billiard-ball atom was the carrier of physical properties of mass, momentum, and energy. These properties were further grounded in the qualities of elasticity, solidity, extension, and inertia of the billiard-ball atom. Physical properties are also dispositional, meaning a physical description is valid at $t + \Delta t$ and is not just valid up until time t and suddenly not. The reason the physical description is valid at $t + \Delta t$ is, again, because it is instantiated in some material, like the billiard-ball atom, which grounds those dispositions. This may seem trivially obvious but a property like momentum covers particulars which fall under the property as its instances, but it is the particular instances which *have* momentum, not the property itself.

In standard physicalism, then, the instantiation is like the simple carrier, or support, for the physical properties, which I will call a "minimal" instantiation. We define a minimal instantiation as whatever satisfies the predicates or relations, having no extra content beyond what is needed to ground them, give them something to inhere in, and to back their dispositional application to future instances. Thus any two minimal instantiations are identical so far as standard physicalism is concerned, because they both successfully instantiate the same physical properties and relations. This is not so in enhanced physicalism, where different instantiations make for different situations or different worlds. In enhanced physicalism of the type I am considering, the instantiation of all physical properties are individualized event particulars in causal–functional relations to each other. These event particulars are individualized manifestations of powers under various circumstances, and are identified by the individual qualities they manifest.

Enhanced physicalist explanations: a priori and a posteriori

To further differentiate the explanatory strategies to be employed in enhanced physicalism, we can introduce a distinction between a priori and a posteriori explanations. My distinction does not concern the metaphysical status of the thesis of physicalism itself (as in Stoljar 2010), rather my distinction between a priori and a posteriori physicalism is a thesis about powers, manifestations, and identity which bears upon explanatory strategies and concepts in enhanced physicalism. In my view, enhanced physicalism should be linked with a posteriori physicalist explanations, not a priori ones.

The difference is best illustrated by an example we have seen before. Imagine some collection of events that go on inside neurons and imagine that each event instantiates measurable physical properties like mass, momentum, and charge. I mean that when we stick a microelectrode *into* the neuron and siphon off the electrochemical energy directly, an event occurs, there is an electrical discharge. Suppose, for example, that we stick microelectrodes into 1,000 neurons, producing 1,000 separate individual events labeled a,b,c, . . . Events produced in this way are Px-level events, described in terms of physical properties. We then configure the individual neurons into various clusters or arrangements, say L-shapes, F-shapes, star-shapes, and so forth. Let the Mx properties be structural properties like being L-shaped, F-shaped, or star-shaped clusters. The events in these clusters will be called configuration-level events. Perhaps one of these configuration-level events is a subject's sensing the color blue. When we say that the sensation of blue is somehow "composed" of neurons firing in some kind of cluster, we mean that the individual configuration event A is somehow to be understood as a set of individual events configured together into a new individual event which we can show with a function symbol, letting $A = f(a, b, c . . .)$. In a priori physicalism, we can actually explain the collective manifestation event A by combining the individual manifestation events a, b, c using function f, so this is just a simple identity of singular terms, of both the powers and their manifestations, with no further caveats. This is what we mean when we say that the event of seeing blue is just the combination of electrochemical discharges we have siphoned off the individual neurons in the cluster with the microelectrodes; all we have to do is reconfigure the individualized manifestation events together to get the collective manifestation event going on in the cluster. Or, once we know all the individual manifestations, and the function f, we know everything we need to know in order to deduce what the collective manifestation will be a priori.

This will *not* be true in a posteriori physicalism. The question which differentiates them is: Is there an *explanation* which starts from the individualized events a, b, c . . ., the fact of their configuration by f (into L-shapes or star shapes), and the basic laws of physics, which would allow us to simply *deduce* from them the collective manifestation event A, the sensation of blue? A priori physicalism holds that the answer is yes.

Examples are a configuration of objects, like matchsticks assembled into an icosahedron, atoms assembled into a ball and stick molecule, or cellular automata constituting high-level behavior that is just the direct result of the instructions given to each individual cell and their initial states (for which see Rosenberg 2004). To give a standard example of a priori physicalism, the higher-order property of "liquidity" can be explained in terms of the basic physical laws governing molecules in configuration and does not represent any new phenomenon arising from the fact of the configuration itself. Once you know enough of the lower-level facts, and the fundamental laws, the rest follows a priori.

In a posteriori physicalism, the way I am defining it, it is still true that in A and in a, b, c ... the *powers* are identical: the powers manifested in a, b, c ... *can* be configured by f into the powers of the individual event manifested in A. Since configuring existing powers cannot add any new powers to the universe, they must be the same in both events. We can thus continue to refer to the same *powers* twice, once under the individual terms a, b, c ... which refer to their individual manifestations in 1,000 separate events on microelectrodes, and again, collectively manifested, under a different individual term A = f(a, b, c ...), which refers to the collective manifestation of the same powers of a, b, c in a different configured event A. Note that we are only equating the powers, and not their manifestations.

On a posteriori physicalism, it is also true that a power is identical to its token manifestations in A and in events a, b, c ... A power is completely present and accounted for in its token manifestations. But it is *not* true that the token manifestation events A and a, b, c are identical to each other, so we cannot write A = f(a, b, c) for the manifestations. They are different, mutually exclusive, events that cannot be equated at all.

As we might expect, when we stick 1,000 microelectrodes into the individual neurons, the sensation of blue *disappears* as that same energy is siphoned off into 1,000 different individualized discharges. When the 1,000 microelectrodes are removed, the sensation of blue suddenly reappears. And when the a, b, c ... are reconfigured by a new function g(a, b, c ...) into a star shape, the blue vanishes and a different sensation, say green, B = g(a, b, c ...) appears instead. Individualized manifestation events a, b, c cannot be configured by f into A = f(a, b, c ...), since those manifestation events A and a, b, c ... cannot co-occur, and if they did, the same powers would be manifested twice.

So, in a posteriori physicalism, the separate manifestation events a, b, c cannot be configured by f to form a compositional, a priori explanation of the combined manifestation event A, even though the powers of A are simply the same powers manifested in a, b, c . . . The powers of a, b, c may even be combined together in f to form A but that doesn't mean the manifestations do.

But now there is a problem, for how can A and f(a, b, c . . .) be identical *qua* powers and non-identical *qua* manifestations? In particular, how can it be the case that the powers are identical with *each* of their token manifestations and even identical *qua* powers across different token manifestation events, but that different token manifestation events are not identical to each other?

The way to understand this is as follows. The same power can manifest different ways across different token events in time and place and still be recognized as the *same* power. This is why we can continue to refer to powers of A as powers of f(a, b, c . . .). Also, each power is *token*-identical with its particular manifestation in a token event. There is only as much power in existence as there are token manifestations of power. There is no distinction between power and the exercise of it, as Hume once put it. There is no such thing as unmanifested energy "in itself," no universal substance or fluid; there are only the actual manifestations of energy in token events across the domains of electromagnetism, gravitation, nuclear forces, etc. Even the potential energy of a particle is still manifested by a small increase in its mass. But particular token manifestations of powers in events are never token-identical to other manifestation events: each manifestation event is separate and unique, and a given power cannot ever manifest in two different token events simultaneously, else it would count double.

For example, consider the energy developed by a water wheel as the gravitational potential energy of the water is converted to kinetic energy of motion and drives a coil in a magnetic field. The same energy is clearly being conserved across these token manifestations and is the same. The energy is also *all present* and accounted for in its token manifestations, either as gravitational potential energy or as the motion of the particles and their force of impact against the wheel, or as the induced EMF on the electrons in the coil.

So the answer to the puzzle must be that several sorts of identity are in play here: (1) the identity of powers with powers, (2) the identity of powers with their token manifestations, and (3) the non-identity of token manifestations with each other. There are always *two* factors needed to identify an event, the power involved and the particular conditions under which the power is manifested, and these two factors are identified differently. Events A and f(a, b, c ...) are identical as powers and non-identical to each other as manifestations. The neural energies are the same in both events, the neural energy is also token-identical, and all present and accounted for in A, when it occurs, and in a, b, c ... when they occur. But the manifestations A and a, b, c ... are not token-identical to each other: when the energy is manifesting as A, the blue sensation, it is not manifesting as the electrical discharges a, b, c ... and vice versa.

So in a posteriori physicalism, we can indeed say that the manifestations of powers in the events a, b, c ... *do not* allow an a priori deduction of the manifestation of those same powers in the event A by configuring them in f(a, b, c ...). The manifestation events in a, b, c and in A are not only not co-occurrent, they are conceptually separate; there is no a priori deduction of the sensation of blue from the 1,000 discharges on the microelectrodes, their configuration in f, and the fundamental laws of physics.

It seems to me that when we discuss the mind–body problem, we are constantly equating the manifestation events a, b, c and A as "the same" merely because the powers are the same and do indeed support an identity *qua* powers ($=_p$), which we can write as $A =_p f(a, b, c ...)$. Yet the actual manifestation events a, b, c ... (the events on the microelectrodes) cannot be configured by f to yield the *manifestation* event A (the seeing of blue). They are not "the same" nor can the manifestations ever co-occur, only the powers can be so configured and that is what $A =_p f(a, b, c ...)$ means, *not* the identity of the token manifestation events.

In science, we identify the same power in two empirically different manifestation events all the time, for example when we say that the kinetic energy driving a water wheel has been converted into electrical energy by driving a coil in a magnet or when the same neural energy is appearing on the 1,000 electrodes or as the sensation of blue. The identical power, the capacity to do work, exists in two different empirical and conceptually separate manifestations. It is the *same* energy, else it would not be conserved, and it is identical with its individual token manifestations because else it would not be all present and accounted for. But because it is conserved, separate manifestation events must remain separate and non co-occurrent, or the same energy would be counted double. Because they are always empirically separate manifestations, there is, therefore, no conceptual or a priori identity between the token manifestations themselves.

The a posteriori physicalist insists on the conceptual separation of two different energy manifestations and will always demand an experimental verification for any two manifestations (such as Joule's experiment). The a priori physicalist, on the other hand, simply assumes energy conservation as a basic law and then equates all of these empirical energy manifestations a priori once he has taken the law for granted. I will show below why I think the law of energy conservation should always be interpreted in line with a posteriori physicalism and cannot be assumed as a premise in a priori physicalist deductions.

The token-identity thesis for powers and their manifestations

Before we go on, I think this is the proper place to head off an objection coming from the theory of powers and their manifestations. In an influential article, Prior, Pargetter, and Jackson (1982) argued that dispositional manifestations are *non-identical* with the powers of their categorical bases. A disposition like solubility is understood by them to be a functional, or second-order property of the categorical molecular properties (powers) of salt and the manifestation conditions of being immersed in water (see also Prior 1985, Mumford 1998). But since the powers in the categorical base cause this manifestation with the help of contingent empirical laws they say the same powers will manifest differently in different possible worlds or situations. Hence the powers are not identical with their manifestations after all.

A strong, and I think correct, rebuttal of this view has been offered under the name of the "token identity theory" of dispositions and bases (Mumford 1998, McKitrick 2003, Heil 2003, ch. 11), which I have been assuming above. According to this theory, the powers in the categorical base are identical with their dispositional manifestations in every *individual* token case, or *particular* manifestation event. Fix the power and the individual circumstances of the power's manifestation, and you fix the token event completely. There is no way this individual manifestation event *could* have been different given the fixing of the two factors. A given power can indeed manifest in a spectrum of different ways empirically, given a *variety* of different manifestation conditions, as Prior *et al.* insist, but still be the same power in each and every one of its separate token manifestations. The token manifestations a, b, c ... *can* and should be used to refer to the powers that produced them, since the token manifestations of powers and powers are identical, and since powers are identical to each other *qua* powers across their different token manifestations. As I suggest, the fact that the token manifestations are not identical to each other *qua* manifestations does nothing to change this. We can and *should* continue to refer to the powers denoted by these terms, even in other situations when they are not manifesting as a, b, c ... but some other way. Hence I can say, as I have been doing, that the same powers are manifested in A $=_p$ f(a, b, c . . .) and B $=_p$ g(a, b, c . . .) as in a, b, c . . .

The composition of causes

Now an a priori physicalist might agree with some of what has been said above and still argue that we can simply assume empirical laws such as energy conservation and the composition of causes as premises in an explanation, and thereby deduce the manifestation event A from the individualized manifestation events in a, b, c configured in f, with the help of such laws. Often enough in science we can deduce the manifestations of the powers of a collection of individuals by configuring the manifestations of the powers of each individual separately. An icosahedron made of matchsticks is a good illustration where the icosahedron's stability and flexibility is entirely explained by geometrically configuring the stiffness and flexibility of the individual matchsticks. There are also many cases in physics where fundamental equations are linear, such that a combination of separate solutions is also a solution, $S(A + B) = S(A) + S(B)$; and there are fundamental superposition laws for vectors and fields.[1]

The problem with a priori physicalism is that this principle of the composition of powers, and manifestations, is an empirical fact which must be guaranteed by experience and tested anew for each kind of case or composition of powers. We can find this empiricist claim articulated by Hume, Mill, and Mach, which runs as follows:

(3) There is no a priori reason why several individual things, with powers which have been observed to manifest in certain ways separately, should manifest in the same way they did individually when we configure them together to a combined manifestation. We cannot reason a priori from their individualized manifestations to their manifestation in combination with each other simply by summing them, as if each were acting separately and independent of the others.

Mach (1883/1960) used this principle as a hammer to demolish a wide range of purported a priori proofs of such principles as the vector super-position of forces and the transitivity of mass. As Mach pointed out, it is

---

[1] There are also of course many cases of **non**-linear systems in nature, such as predator–prey differential equations; and patterns of firing in neural clusters in fact are especially strong candidates for modeling as non-linear systems.

not a priori true for example that if the mass of a is to the mass of b as 1:1 and the mass of b is to the mass of c as 1:2 then the mass of a is to the mass of c as 1:2. Experiment alone justifies the extension of the mass principle to all bodies and configurations of bodies. Today physics textbooks follow Mach and state all compositional principles as important empirical propositions, *not* mathematical theorems. The principle of vector addition is a law of mathematics but that forces should combine like vectors is an empirical fact.

Notice Mach did not prove that the a priori deductions were invalid, leading from true premises to a false conclusion, but that they were circular because they assumed premise (3) in some form or other, rendering them vacuous. If an a priori physicalist says, for example, that the empirical fact of combination stated in (3) is to be added as a *premise* to his derivation, then this circularity objection may also be urged against him, since the combined manifestation must be mentioned explicitly in the principle itself, before "deducing" the combined manifestation from the individual ones with the help of the principle, so any such deduction is vacuous. On the other hand, if the a priori physicalist is allowed to pack the premises with all the empirical data and laws that he wishes, certainly he can make it seem that a priori physicalism is always true, but in this totally empty and circular sense of certainty that Mach effectively exploded.

In relation to neutral monism, the same point has been made effectively by Michael Lockwood, albeit in a different context. Lockwood was criticizing a certain kind of panpsychist Russellian physicalism (to be considered below) where there are a priori "deductions" or "constructions" of conscious phenomena like sensation from little "proto-sensations" in neurons or atoms.

> Their [basic dispositions of matter] potential for generating awareness could be a matter of the application of certain currently unknown laws to their familiar physical attributes (in which laws, of course, there would be an essential reference to the emergent attributes). This fairly elementary point seems to have escaped those authors who have argued that, if we are made of electrons, quarks, gluons, and the like, then—given that we are conscious—electrons, quarks and so forth must themselves be possessed of some sort of primitive proto-consciousness. As I see it, this is a complete non-sequitur. (1993, p. 281)

According to Lockwood, the combined activity of neurons and their ability to manifest sensations when configured should be understood as a case of an empirically "emergent" manifestation which cannot simply be deduced a priori from the available information about individual neurons, their

configuration, and the physical laws governing those individualized manifestation events. Unlike, say, the account of liquidity, in which the
property of "liquidity" need never be mentioned in the fundamental laws
used to derive it from more basic phenomena, the empirical laws we need
must mention the higher-order manifestations explicitly in the laws
themselves. So these laws *cannot* be couched solely in terms of lower-
order events, fundamental physical laws, and micro-configurations. And
if those laws which explicitly mention the composition of manifestations
to a configured effect are simply assumed as premises, then we then have a
totally vacuous explanation of those combined manifestations, as Mach
predicted. Lockwood's concept of an "emergent" law is also to be found in
J. S. Mill, who called them "heteropathic" laws, and in C. D. Broad, who
called them "trans-ordinal laws" (see McLaughlin 1992, pp. 51, 61, 80–81).
I object to the term "emergent" and would prefer to leave it out of my
discussion, using only the vocabulary of powers and lower- and higher-
order manifestations, to be explained in what follows. I see higher-order
manifestations as empirically *separate* events from lower-order ones, not as
"emerging" from them.

Russell's hypothesis redux

Let us now return to our reconstruction of Russell's hypothesis. According
to the way I phrased it, Russellian event particulars are expressions of
causal powers manifested directly in events. These manifestations are the
event's individual qualities. There is no mystery about qualities, nor are
they mental; they are simply the concretely and directly manifested effects
of natural powers in plain language and they occur around us all the time
in nature, in both observed and unobserved events. The quality of the
particular event *is* the concrete manifestation of that power under the given
circumstance in which the event occurs. These natural Russellian "event
particulars" instantiate, or ground, an externally described network of
physical objects, properties, and relations. No further grounding is necessary, nor can there really be anything further behind the manifested
qualities of an event, such as further substances or objects, not even
unmanifested powers. I do not believe anyone can give a real example of
an unmanifested power in nature, since all natural energies are completely
present and accounted for in their token manifestations (even potential
energy). It is true that powers currently manifesting in one way could
certainly manifest another way later, or would be doing so now, if the
circumstances of manifestation had been different, but these would be the

same powers manifested in token events now, manifesting differently in some other way, or in some other circumstance, *not* presently unmanifested powers ready to appear from nowhere (as in Martin 1993 if I read him correctly).

It has been often objected that powers cannot serve to ground an ontology because they are by their very nature "ungrounded" and simply devolve upon further powers in an infinite regress (see Bird 2007). But this is clearly untrue if powers *are* concretely grounded in the qualities they manifest and with which they are token-identical. This token identity stops the regress. Secondly, it has been objected that powers are like the "virtus dormitiva" ridiculed by Molière for explaining the powers of the sleeping drug with a sleep-inducing power, when clearly more can be said by grounding the powers of the sleeping drug in its categorical chemical composition and how it acts dispositionally upon the brain. But at a still more fundamental level, these categorical chemical and atomic structural explanations give way again to potentials and forces and events. The only reason the atoms and molecules have the categorical shape and form they do is because of the more fundamental powers of their constituent parts. A molecule is structured like a tetrahedron because the powers of atoms conspire to act in a "tetrahedral way," and so on down. Hence fundamental powers and their manifestations in events may well be the ultimates of physical explanation after all, not traditional categorical properties like shape and size. Instead of matter with categorical shape and size why not force? I believe that what many people conceive visually as solid objects are really powers on a closer and better-informed view. This underlying network of natural powers and their manifestation in forces provides the same generality and objectivity to our view of nature as a matter-based view and I see no a priori reason why we should prefer the latter.

Finally philosophers often argue that mere individual events require objects or universals in order to reidentify them. But according to Russell's hypothesis each individual event can be identified by its qualities and causal relations, which serve to pick it out uniquely. Russell insisted on such a principle in a later article on the "Principle of Individuation" of events (1947/1997, pp. 242–243). We do not need objects or properties or even repeating event-types to identify event particulars one by one and therefore across times and places or even if they only occur once in the entire history of the universe. Nor are event particulars so-called "bare particulars" on this view, since they come fully "clothed" in their own individual qualities.

But Russellian neutral event particulars are *so* physicalistic in nature that there does not seem to be any reason to assume that these natural qualities in physics have anything at all in common with our sensations, which are qualities of a very different order. A typical event particular in Russell's universe would be something perhaps like the emission of a quantum of radiation from an atom, having little or nothing to do with events in the human nervous system at a very different scale of complexity and size. So what is the connection? How does one *go* from these qualitative expressions of power in physical events like electron jumps in the atom to the manifested qualities of our sensations like blue or green? That Russell believed there was a relationship between natural qualities and those of our sensations there can be no doubt, but he did not provide any further details on what that relationship was. Hence many contemporary philosophers of mind (Chalmers 2002, Strawson 2006) have been tempted to see Russell as a "compositional panpsychist," or to give his hypothesis that kind of interpretation where little proto-qualities in matter combine into the qualities of our sensations by a construction based on a line of descent from physical proto-qualities to sensation qualities. Myself, I believe there are much better options, more in line with the historical Russell and the realistic direction in which he was headed. My a posteriori enhanced physicalist view may seem more complicated but I believe it is ultimately the more parsimonious option.

Micro versus macro manifestation events

Let me now show how I think Russell's hypothesis applies to the problem of explaining sensations in an enhanced physicalist way. The case I want to consider is again one in which we have a base micro-level, say, of event particulars in neurons, where their powers are individually manifested in events a, b, c and d, e, f. Consider again sticking 1,000 microelectrodes into individual neurons and siphoning off the energy of the cells as they fire. The sensation vanishes from consciousness and the physical events are recorded. Each such event a, b, c ... satisfies an individual physical description P. We then remove the 1,000 microelectrodes and allow the sensation to reappear. In this separate event, these same *powers* of neurons are configured together in a cluster via some function f to an individual configuration-level, or macro event $A = f(1,000 \text{ neurons})$ with its own differently manifested powers.

Now, consider the configured events A and B in Figure 5.1. Let these individual configurations be assembled by the functions $f(x, y, z)$, $g(x, y, z)$

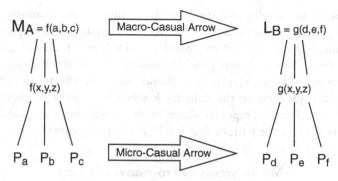

Fig 5.1   Macro- and micro-causal relations

and described by the macro-properties they instantiate, such as M-shaped, L-shaped, star-shaped, etc.

If the combination is an a posteriori physicalist one, the concrete manifestations of the configured powers are simply the manifested qualities of the configured events A and B, like blue or green. These higher-order qualities are the concrete manifestations of the powers of higher-order *configured* events. We will then say that the configured powers form their own higher-order macro-causal network and that the qualities that instantiate that network form their own higher-order "quality space," such as the familiar phenomenal spaces of color or sound. The manifested qualities of the M-shaped or L-shaped configurations are *not* deducible a priori, by a line of descent, or by relations of similarity, or by other conceptual relations, from the qualities of the individual manifestations of the powers at the base level. These events are simply a different *kind* of qualitative manifestation of the basic powers in the cells, when they are configured to manifest those powers in a collective way and not in individualized ways. I would have no objection, for example, to siphoning off the energies of 1,000 neurons separately, guiding them along wires, then reconfiguring these energies by allowing them to act upon each other in a combined manifestation of A again. I cannot see how this manifestation could fail to be red or green.

On enhanced physicalism, our experienced sensations manifest in their qualities the *configured* powers of our individual neurons, or rather the complex electrochemical events in them, and the macro-causal relations among these configurations in quality space. A star-shaped event might be instantiated by the quality blue, L shapes might instantiate the quality green. I think we actually experience these configured events in our brains as colors, tastes, smells, and the like, depending on the anatomic region and the pattern of macro-causal relations there among the configured cells (see Kandel *et al.* 2000). Configured L-shaped and F-shaped and star-shaped events in our brains probably just *feel* to us like the qualities of colors, sounds, touches, etc. (see Banks 2010, p. 185 n. 17). Perhaps Democritus was even on the right track when he thought a "sharp" taste or smell was due to the pointy shape of the atoms that composed them, only perhaps it is the pattern that is felt as sharp or pointy.

Micro- versus macro-causal relations

The problem of relating sensation qualities to the ordinary qualities manifested in individual physical events seems to me to reduce finally to a question in the philosophy of science about the viability of robust macro-causation among macro-events, and in macro-causal networks, versus micro-causation among individualized events (see Banks 2010). So we must turn, ultimately, to issues having to do with macro- and micro-causation to answer our question of what *kind* of natural events sensations really are and where they fit into the puzzle of the physical world.

Most philosophers will admit the idea that although configuring powers cannot introduce any additional or novel powers into the universe on pain of violating conservation laws, a configuration of individuals will still possess differently *manifested* powers that the individualized powers or parts do not manifest in isolation. Certainly no extra power or energy is introduced into the universe simply by configuring micro-causal powers in certain ways, but *how* we configure those powers will certainly change the way the powers are manifested. Configurations channel the manifestations of powers: they "block," "constrain," and "induce" these manifestations in different ways. They also induce, constrain, or block the occurrence of other configurations. A configuration of atoms does not kick, per se; the individual atoms and their powers do all of the kicking, but a configuration controls the *circumstances* of how the kicks can manifest their effects collectively and thus configurations play a major role in actually determining the manifestation even of individual powers of atoms. Higher-level facts about configurations determine the circumstances, and thus the manifestation conditions, for lower-order powers in certain ways or channels. A better way to say this is: configuration-level powers manifest the powers of lower-order individuals *in the way in which they are configured*, which cannot be left out of account, as this is just as important to the circumstances of manifestation as any other facts. We can then call these higher-order relations, which structure the manifestation conditions of powers in configuration, "macro-causation." A similar distinction is made in Dretske (1993) between "triggering causes" and the background "structuring causes" of events, and by the many authors who have dealt with this subject. I claim no originality here, I would only emphasize that the sort of macro-causal powers I have in mind, directly affecting the *circumstances* of manifestation, are also *real* powers and not just higher-order explanatory heuristics. Higher-order powers are not just effective explanatory devices, or epistemic concepts; they are causally effective and *that* is why they are useful in giving explanations, not vice versa.

Common objections to macro-causation

There are three common objections to the idea of macro-causal networks:

1. Macro-causal powers and networks are derivative of micro-causal relations, which realize them, supersede them, or screen them off (Epiphenomenalism).
2. Macro-causal networks overdetermine events by giving two redundant causes for the same events (Kim's Exclusion argument).
3. Macro-causal networks violate the principle of the conservation of energy (the Causal Closure or Double Counting objection).

I don't think any of these objections rule out macro-causal networks of the sort we are considering. On issue 1, the 1,000 individualized events of siphoning off the energy of discharging neurons and the individual experience of the sensation of blue are mutually exclusive. If a person is sensing blue and we siphon off the energy somewhere, the sensation disappears and the energy shows up on the external measuring device instead. The same energy is manifested two different ways in two different events, but never at the same time, so a fortiori one event cannot be said to supersede, pre-empt, or screen-off another, except trivially, since they could never have taken one another's places to begin with.

Must macro-causal relations among configured events be weaker, nomologically, than micro-causal relations? Must they be "screened off" by stronger lower-order causal relations? As I have argued previously (Banks 2010, p. 179 and n. 13), we may not actually have a strong micro-law that individuals in a *particular* configuration $M_A = f(a, b, c)$ are likely to induce, constrain, or block the powers of individuals in a particular configuration $L_B = g(d, e, f)$ but we may possess a stronger macro-law that configurations of *type* M are likely to induce, constrain, or block configurations of type L. For example, the behavior of individual neurons may well be random, non-linear, or even chaotic, responding to local electrochemical conditions, but the behavior of a configured region of neurons responding to evolutionary or environmental pressures at a macro-level may be far more stable and law-like. Nor does the presence of a higher-order macro-law necessarily indicate a lower-order "screener off" for the correlation (for that view see Papineau 1993). We may not be able to find such a condition C holding strictly of individual powers and their expressions at the lower level that we can conditionalize upon and which screens off the correlation between the macro-powers and their manifest-ation events. I think this is especially likely to be true of neural events.

On issue 2, Kim (1998, 2005) presents his well-known "exclusion argument," which he thinks rules out the macro-causal arrow in Figure 5.1. Kim says that when a configuration M changes to a configuration L, this is entirely due to the fact that micro-level individuals have kicked each other individually in such a way that a new configuration results. If we attribute the cause of the configuration to *both* the individual powers acting at the micro-level and a macro-level cause, such as the way they manifest in their configuration or arrangement, Kim thinks that there will be two causes for the configuration L and that this indicates a problem of overdetermin-ation. I will simply echo Kim's critics here in asking: Why? Why can't events be co-determined by micro-level powers acting in macro-level manifestation conditions, which co-determine and effectively channel how those powers *manifest*? Both seem *sine qua non*.

I think Kim does recognize that configurations of individuals and their powers must be taken into account, but he offers to reduce macro-level configurations to micro-level configurations and pack it all into the supervenience base. We saw before that this runs afoul of the empirical principle of the composition of causes, if the higher-order manifestations must be mentioned in the supervenience base, as they always must when powers and manifestations are at issue. Kim says (2005, ch. 2) that we can always realize macro-level facts and arrangements in successively lower levels of organization so that macro-level configurations of individuals are successively realized in levels all the way down to the elementary particles and forces. We thus end up with a bizarre "baseball-sized" configuration of fundamental particles at the lowest orders of matter and energy. Why is this? Kim's procedure is top down, not bottom up, and identifies the configuration and its powers at the macro-level, not building them from the micro-level up as we would expect. He cannot say of a micro-configuration whether it is a "baseball" or not, alone, without the cascade of realizations starting from the top, so he is implicitly using the higher-order concept in his supervenience base. And if enough micro-level individuals are configured and these relations are spread out to their proper scale, the configuration *ipso facto* grows into a macro-level fact needing description in macro-level terminology anyway. As I said, I think Mill, Hume, Mach, Lockwood *et al.* have already effectively countered this sort of a priori physicalist reductionism by characterizing a class of empirical laws, the composition of causes, in which the mention of higher-order phenomena cannot be eliminated in the description of how lower-order stuff manifests its combined powers.

Finally, we consider the causal closure and double-counting issues. In some ways, the closure argument is based upon a widespread misunderstanding of the law of the conservation of energy in physics (see Banks 2010, p. 178). As we already saw in the discussion of Mach, the law states that in a closed system, the energy of the initial state of a process is equal to the energy of the final state of the process, but it does *not* specify anything about the way the energy changes form, manifests, or what path it takes between those states. Indeed it does not even predict that a change will take place at all, since a body could hang in the air and still satisfy the principle of energy conservation. Also, the conservation of energy does not rule out the formations of higher-order configurations and causal networks, so long as they do not contribute any extra energy to the universe, but merely channel the available energy in various ways, forcing it into different manifestations, which we may call macro-causal relations. The conservation law says nothing for, or against, the existence of macro-causal powers, events, or networks.

The alleged double counting of powers at the micro- and macro-level is a more interesting problem. Granted that no extra power is introduced into the universe by configuring the powers of individuals to a combined effect, how can there be anything for macro-causal relations to *do*, since they seem to lack powers of their own? On the other hand, if they have powers due to their configuration, over and above the powers of the configured individuals, then they introduce extra powers into the universe and that does violate the principle of the conservation of energy, in letter and in spirit.

The issue can be resolved by insisting, once again, on the distinction between a power and its manifestations. Powers can be token-identical with their manifestations in any event and can even be identified as the same power across *different* token manifestation events. But even though the events a, b, c ... , A, and B denote the same identical powers manifested three different ways, they cannot be "set equal" to each other *qua* manifestations. These are different terms for different events which cannot co-occur and which are, in fact, mutually exclusive, else the same power would indeed count double or even triple, since it would be manifesting three ways at once. But it never does. The same neural energy that collectively manifests as the sensation of blue can be siphoned off into each individual cell, at which point the sensation disappears just as the readings appear on the individual 1,000 microelectrodes, but both manifestations *never* occur simultaneously so there is never any double counting of powers or effects.

But are the powers really identical across different manifestations? A reviewer of this chapter (as an article) gave the example of butane and isobutane, two different configurations of the same hydrogen and carbon atoms with the same individualized powers, but with apparently very different chemical macro-powers when configured into the two isomer molecules. If the underlying powers are "the same" in the atoms *and* in the two configurations, why does configuring them differently make the powers manifest differently at the configurational level? The answer in short is that when the powers a, b, c are configured, by f and by g, they manifest differently in the two molecules f(a, b, c) = A butane; g(a, b, c) = B isobutane. The powers of A and B are indeed the same but the configurations are not, so the individual manifestation events A and B are not the same either. If we want to express the sameness of the powers in B and A, we can still do so by including the reconfiguration of the powers, r, and writing $r(A) = B$; $r^{-1}(B) = A$.

Another reviewer mildly complained that the foregoing account is not an "explanation" of the mind–body relation, since it does not show how electrochemical energy, manifesting in 1,000 neurons separately, "assemble" to manifest collectively as a unified sensation quality like blue or green. This of course assumes that compositional, a priori physicalist explanations are the only acceptable ones, which I would dispute. On my account, the qualities manifested by macro-events like blue are not literally "composed" or "assembled" out of the micro-level manifestations, say in 1,000 spikes in neurons. The events are mutually exclusive, not co-occurrent. Also, the configured events have their own higher-order qualities manifested by configured powers, like star-shaped, L-shaped, and M-shaped events. The macro-qualities and macro-causal relations first arise *there*, at that level, and are not assembled out of different quality manifestations at a lower level, as for example in compositional panpsychism, of which more below. Suppose a "star-shaped" cluster of neurons possesses a distinct firing pattern, but only *as a system*, perhaps linear, perhaps non-linear. It could also be that the powers or potentials which manifest in events of the sudden firing of neurons, when configured differently, will induce a different, coordinated pattern of firing, or that the same powers configured differently manifest as different events. Consider that these repeating star-shaped patterns of firing could be what we sense as the manifested sensation quality. The pattern is where the quality first appears, where the star shape is first formed, not inside, not below, not at a greater level of detail. The manifest quality of the sensation is not "assembled" out of the natural qualities manifested by individual neurons, nor does it even "emerge" from them as Lockwood thinks, they are simply two separate, empirically *different*, manifestation events of the same powers which need not be conceptually related at all, echoing Hume's argument about the conceptual independence of "original existences," against the a priori.

While my account is not therefore a compositional, nor an emergentist, account, it is at least a *unified* neutral monist account. I would accept ontological unification as an explanation in lieu of an a priori physicalist compositional account or wild, *sui generis* emergent properties. We still have a physicalistic quality manifestation of some kind on both sides to cement the underlying identity between powers of sensation qualities on one side and powers manifesting in other natural physical qualities on the other. But unlike other such materialist a posteriori identities, we do not unite concepts of radically different *kinds* of things, pain sensations and c-fibers, appearing on either side of the identity. In those kinds of identities, two radically different kinds of thing are pasted together and asserted to be identical, without giving us the slightest reason to believe it, for example an identity between the world of particle physics on the one side and experienced sensation qualities on the other which does no explanatory work at all. In cases like this, we may question why the relation should be one of identity at all. Why not dualist interactionalism, or psychophysical parallelism, or even pre-established harmony?

Finally, I have been asked if my view isn't just like the British Emergentism of C. D. Broad and others (for which see Brian McLaughlin's excellent 1992 overview). According to McLaughlin, an emergentist is one who believes the world is made up of atoms and molecules with their micro-level powers, but who also believes that when these atoms and molecules are configured in certain ways, additional novel *powers* of the configuration result which could not have been predicted, or which were not even there in the constituent parts until they came together in the configuration (1992, pp. 50–51). This is certainly not the case in enhanced physicalism, where the powers are the same in lower-level and in higher-level configured events; only the token manifestations of the powers are different. Second, unlike the British Emergentists who speak about *objects* and configurations of objects, we have been speaking about powers and their manifestations as the two different factors in producing *events*. I am aware of no such distinction in British Emergentism. The emergent "heteropathic" or "trans-ordinal" laws mentioned above in the discussion of Lockwood, Mill, and the Emergentists do indeed seem to me to be a common feature of both positions. But for me, the presence of those laws is necessary for giving a posteriori physicalist *explanations*, having nothing to do with the metaphysical emergence of "novel powers" of wholes not possessed by the parts.

Russell's hypothesis, and the a posteriori enhanced physicalism that I think best extends it, are thus to be recommended over standard physicalism on grounds of parsimony and ontological unity, in the sense that enhanced physicalism preserves all of the results of physics and still makes room for the reality of sensation and human experience. Moreover, enhanced physicalism does so in a *physicalistic* way, by enriching our understanding of powers, manifestations, and the different orders of causation that operate in the natural world. The phenomenon of sensation simply falls into place as a certain type of physical event among others in nature. The separate category of mental phenomena simply ceases to exist, except as a provisional way of talking.

To summarize my results, in enhanced a posteriori physicalism, there are two issues to be distinguished: (1) ontological monism and (2) explanatory dualism, or:

1. What it is for one manifestation of powers collectively to be different from a manifestation of the same powers individually. I do not believe there are any fundamental differences in kind between the event of seeing a blue patch and the event of having all the *configured* neurons fire in the region of the brain responsible for seeing the blue patch. The quality blue and the individual electrical discharges are just different and mutually exclusive manifestations of the *same natural powers* which we mistakenly see as belonging to totally different categories of event.

2. What it is to *explain* or equate one manifestation of powers collectively in terms of the individualized manifestations of the same powers in a *different* set of events. No one could explain one kind of natural

manifestation by citing another empirically different manifestation, especially when the two events cannot co-occur. For example, it is not clear to me that we "explain" the manifestation of the powers of many neurons collectively when a person sees a blue patch by pointing to the different manifestations of those powers individualized neuron-by-neuron and interfered with directly by siphoning off the energies into 1,000 microelectrodes, or by making an indirect scan or recording of the brain region that is active when the person is seeing the blue patch. That is not an "explanation": the events are different and do not even co-occur. I do not think we can ever simply "add up" the individual manifestations neuron-by-neuron and hope they will configure into the collective manifestation of those same powers in our sensations, because these are different and empirically separate events: either the one happens or the other one does, not both. I am well aware that many philosophers of mind have defended an a posteriori physicalist view, involving ontological monism and explanatory dualism, as here, but those views are predicated on a materialist monism, not on neutral monism, with its simpler ontology and greater economy.

Other neo-Russellian views: Chalmers and Strawson

The view I have developed here comes from my study of the classic neutral monism of Mach, James, and Russell as an historian and philosopher of science, and not from the contemporary revolt against physicalism under way in the philosophy of mind. As such, my view was never *intended* to fit with reconstructions of Russell's hypothesis originating in philosophy of mind, based upon different intuitions and thought experiments. I have only one experiment about the mutual disappearance of blue, and the appearance of discharges on the electrodes, which is not a thought experiment. Nevertheless, I do want to consider two of these reconstructions of Russell's hypothesis by David Chalmers and Galen Strawson and say why I prefer my own. Chalmers presents a view he calls Type F-monism, which he summarizes as follows:

> Type F monism is the view that consciousness is constituted by the intrinsic properties of fundamental physical entities: that is, by the categorical bases of fundamental physical dispositions. On this view, phenomenal or proto-phenomenal properties are located at the fundamental level of physical reality and in a certain sense, underlie physical reality itself ... As a bonus this view is perfectly compatible with the causal closure of the microphysical, and indeed with existing physical laws. The view can retain the

structure of physical theory as it already exists; it simply supplements this structure with an intrinsic nature ... (proto)phenomenal properties serve as the ultimate categorical basis of all physical causation. (Chalmers 2002, p. 265)

Chalmers sees the categorical basis of dispositional physical properties as "protophenomenal" qualities, embodying the causal powers to instantiate a network of events and relations in physics. However, he rejects the idea of a higher-order causal network:

> If the low-level network is causally closed and the high level intrinsic properties are not constituted by low-level intrinsic properties, the high-level intrinsic properties will be epiphenomenal. The only way to embrace this position [i.e., of a higher-order causal network in which higher-order qualities inhere—E.B.] would seem to be in combination with a denial of microphysical causal closure, holding that there are fundamental dispositions above the microphysical level which have phenomenal properties as their grounds. (2002, p. 267)

There are then two options: accept epiphenomenalism for higher-order qualities, making our sensations into epiphenomenal qualia, or accept some form of compositional panpsychist theory where the higher-order qualities of our sensations are assembled out of the lower-order protophenomenal qualities. This would be a form of a priori physicalism.

In my view (and I think also Stoljar's 2001) we can already see that epiphenomenalism about our sensations is false, because everyone admits that phenomenal sensation qualities *appear* in experience. But anything that appears must be manifesting some kind of effect on its environment, else it would not appear anywhere at all. Sometimes that thesis is traced back to Plato's *Sophist* when the Eleatic Stranger says that "the definition of being is simply power" (Jowett 1871).

To give it a name, call the macro-causal role that sensations instantiate in the brain their phenomenal "appearing" role. Properly understood, on enhanced physicalism (though not on standard physicalism) cases like the inverted spectrum are absurd, because what they ask for is impossible. We are asked, for example, to hold *all* causal and dispositional roles of brain and behavior fixed and contemplate a situation in which the phenomenology were different, switching ROYGBIV for VIBGYOR. But if we hold the appearing role fixed, as one among these physical causal roles, and thereby "lock in" the instantiation of ROYGBIV, then the phenomenology cannot vary, contrary to hypothesis, and if we do *not* hold it fixed then we do not really, and truly, fix *all* causal roles and this premise of the thought experiment is false.

Kim's proposal to strip away functionally just the functionally defined "similarities and differences" of sensory qualities from their phenomenology (Kim 2005) is just as misguided on enhanced physicalism. Those phenomenological similarities and differences whose causal roles mesh with behavior and physics are not "functionally abstracted" from phenomenology: they *are* the appearing roles of the phenomenological contents and cannot be separated at all from the rest of the causal, or functional, structure. The quality instantiates the causal, or functional, appearing role it plays in the mesh continuous with the rest of the physical world. Ignoring a real physical phenomenon like the appearing role is not physicalism at all.

The second option, then, is to think of higher-order qualities as "composed" somehow out of lower-order protophenomenal qualities. I think this is very unlikely to be true. First, it would involve some a priori line of descent or similarity between human sensations and protophenomenal sensations in matter, a strained analogy given the vastly different orders of complexity between say the powers of a particle, and the qualities they would manifest, and a tangled city of neurons in the brain and the rich sensation qualities they plainly manifest there. If there is any qualitative similarity it is likely to be so general as to be empty—as we might say of *any* two things that they are similar in some respect or other, with no explanatory value to the comparison. Furthermore, and this is perhaps a stronger argument, qualities are empirical manifestations which cannot be anticipated a priori. No one can predict the taste of fruit or an unexperienced color. So it is unlikely that any philosopher's a priori deduction or assemblage of higher-order qualities from protophenomenal qualities could ever succeed, even in principle. It is better to treat each separate empirical manifestation of a quality as different and conceptually independent.

A final reason for rejecting protophenomenal qualities has to do with evolution. Sensory manifolds and their qualities are clearly evolutionary adaptations to differentiate a macroscopic environment of light reflectances (the three-cone system) and the struggle to maintain the constancy of colors across different absolute luminance levels (the red–green and yellow–blue inhibition system). To think of these sensory manifolds of color as coming together via some "compositional conspiracy" among proto-phenomenal qualities, to eventually, billions of years later, form blue at the macro-level out of proto-blue at the micro-level, just strikes me as bizarre. It is much more likely that this macro-causal role among neurons is indeed instantiated by the sensation blue but that individual neurons do not exhibit anything like proto-blue. The blue appears when the *configuration* of neurons evolves a pattern to manifest it and it simply did not exist before. I agree with Lockwood that extending sensory qualities by a line of similarity or composition from our human sensations all the way to neurons and atoms seems absurd and unfounded in the extreme. Anyway a true "compositional" theory would *start* from the bottom up, from the simplest physical qualities, and not infer from the top down, from our sensations of blue on down to proto-blue. Blue should rather be called "proto-something else" if the theory were serious.

Galen Strawson (2006) takes an even odder view. He starts out by saying that any true realist or materialist must accept the reality of sensations alongside physical objects and events, which I think is true, but then he veers off in a panpsychist direction. Strawson believes that sensory qualities cannot "emerge in a brute way" from non-sensory materials like neurons and atoms. Hence there are qualities similar to our sensations clear on down to the smallest parts of matter. Likewise, he believes that sensory qualities cannot occur except in an ego who beholds them. So, in addition to protophenomenal qualities in atoms, there are little proto-egos all the way down too. The temptation might be to dismiss this view as "turtles all the way down," but I think we can do a better job by contesting its premises. Strawson's "no emergence" argument is based upon an uncritical acceptance of a priori physicalism that results from ignoring the composition of causes. Thus one is led to assert that there is nothing in the whole or configuration that is not in the parts. I would agree for the powers but not for their manifestations, where the macro-configuration does make a real difference to the way the individualized powers are able to manifest collectively. Hence, to rephrase, there is nothing in the manifestation of powers in a configuration that is not in the individual manifestations of the powers *except the configuration itself.*

Conclusion

So, however mysterious the mind–body problem may be for us, we should always remember that it is a solved problem for nature. All we have to do is figure out that solution by naturalistic means. My reconstruction of Russell's hypothesis is one among many, and not the easiest view to understand or explain; but mental economy is not always an indication of truth, and the view I have presented here seems to me much more physicalistic in its bones than any of the other options presented so far, especially the panpsychist ones. For me, Russellian qualities are simply empirically manifested natural powers, micro and macro, with nothing ultimately mysterious or mentalistic about them.

CHAPTER 6

The problem of extension: a constructivist program

Introduction

A major problem for the realistic empiricist view is how to construct a world of *extended* objects and measurable properties of systems out of elementary events, to create a kind of general manifold for physics. As we saw, Russell's ambitious attempt to construct space-time out of quality overlaps in the *Analysis of Matter* suffered from a circularity, since he was forced to assume a background extension, so that his construction amounted to no more than the construction of a special kind of extension *within* another extension. A similar objection could also be raised against Mach. For all his talk of eliminating "psychological" or "metaphysical" space and time from physics, he sought to "relativize" spatio-temporal motions and laws, on the way to eliminating the spatio-temporal format, but he never actually explicitly constructed or eliminated extended representations from physics, since all of the relations he assumed were also extended ones. Even Mach's Principle would not eliminate extension and motion from physics, it would only relativize them (see Banks 2003, ch. 12).

In this chapter, I will examine a different constructivist tradition, which stretches from Leibniz to Herbart, Riemann, and Grassmann. In this tradition, extension is *not* a fundamental property of nature, nor an intuition that must be taken for granted in geometric thought or in physical theories of nature. Rather, extension can be constructed in an abstract combinatorial fashion *without* assuming an extended background drafting board. This tradition, I believe, reaches its intellectual apex in Grassmann algebra, first presented in the *Ausdehnungslehre*, which is not always recognized as a philosophical construction of space, as it was originally intended. Having laid out the ideas of the constructivist tradition, I will adopt Grassmann algebra (in its modern form) as the proper language for the construction of extension in realistic empiricism, starting with the "point-event" as the basic entity of interest and building up other extensions in a purely algebraic manner. As it turns out, many of the features of realistic empiricism can be captured formally by Grassmann algebra, for example the idea that a point-event is an individual manifest-ation of power, or potential, and the perspectival structure of events. The mathematical details will be minimized here, but the interested reader may want to consult my 2008, 2013c articles for more details.

The philosophical problem of extension

Everything is extended. Extension in space and time is probably the most general property of nature, as Descartes pointed out. But is it a simple, fundamental property that must always be taken for granted, or is it constructed out of still more fundamental properties or relations? Des-cartes claimed, along with most philosophers and scientists, that extension could neither be explained nor constructed. Leibniz denied this and insisted that extension indicated internal complexity or structure to be analyzed. Followers of Leibniz, such as Wolff and later Herbart, sought an explicit construction of extension from unextended points and instantan-eous forces in a physical monadology. Finally, the mathematicians Riemann and Grassmann took a turn at developing extension from scratch from associative and dissociative processes.

The first problem is to define extension in a non-circular way, without assuming it. We think we can point to a row of dots or an extended length or duration and say that the dots are all separated and "apart" from each other. Similarly, what definitions there are of extended magnitudes, from the medievals to Kant, talk about the *apartness* of parts of magnitudes like a length or a time, the parts of an object, stages of a process, and so forth. The opposite is an intensive magnitude with a degree or intensity, but no extended parts. These definitions are clearly circular, since the "apartness" or "insideness" in question simply assumes the intuition we are trying to define. Even divisibility assumes extension, because the divisions are made within an extended magnitude.

The next thought is that there must already be a definition or analysis of extension somewhere in mathematics, in the construction of the natural or the real numbers or in topology. But a length and a degree, say of temperature or density, are both representable in the same way by the real numbers. At no point is any appeal made to the intuition of "outside-ness" or "inside-ness" of parts or degrees in the construction of the reals. The intuition of extension plays *no role* in the abstract construction of the reals out of sets; it rather has to do with how we *represent* the real continuum. Mathematics makes no distinction whatever between an extended representation and any other interpretation of its structure. You could think of sets as points collected in extended space, or as anything else you wish. And nowhere in mathematics is it ever specified what would make for an extended interpretation of numbers or points, not even in topology; it is simply left as an issue of which intuitive representation one chooses. When a specific space-time structure is built up, we assume a blob-like extension of points, call it $R \times R \times R \times R = R^4$. We then define a topology for the blob (an open ball, a sphere with boundary, a torus) by laying constraints on neighborhoods around points, or by shrinking extended regions. A coordinate mesh can be imposed, the intrinsic curvature and the metric can be defined, all the way up to the causal structure of lightcones in the space, but nowhere has the assumption of the extended blob been justified as opposed to any other interpretation of the sets making up R^4. We are forced to conclude that extension, just like sense or order,[1] is not defined in mathematics, rather it is an intuition we bring to mathematics when we make an extended representation of its structure.

This brings us to Kant's view in the *Critique of Pure Reason*, that the extended "drafting board" of intuition, on which all geometric or physical constructions are done, is necessarily *assumed* and cannot be analyzed, or brought under simpler concepts. Kant points out that things are contained in space but not by *falling under the concept of space*, the way that individual horses fall under the general concept or definition of horse. Kant did not believe there was an explicit concept of extension capable of explaining in what respect objects or regions are spatial. Instead, he was content to assume a kind of general undifferentiated extension as a basic intuition, whether in geometry or in physics. In the *Transcendental Analytic*, Kant constructs the "space of the understanding" and the objects in it by an explicit synthesis, such as drawing out a row of dots or a line in space, but he still insists the prior undifferentiated extension is always presupposed and is somehow impervious to the analysis of the understanding and its

[1] Directions, such as right and left, have to be given by some means external to mathematics, if we are to avoid begging the question. Order can be represented, say by odd and even permutations, if we *already* understand the order 123 in the positive sense, i.e., the handedness of a coordinate system, where one stands in the origin and counts the axes off in clockwise (right-handed) or counterclockwise (left-handed) fashion. Similarly, the ordered pair $\langle u, v \rangle$ can be defined as being in the *same* order as $\langle x, y \rangle = \langle u, v \rangle$ if one *already* understands what the order of $\langle x, y \rangle$ is. Russell (1903) proposes including order or sense among the undefined notions, but says nothing about extension.

discursive concepts. Likewise, in the *Amphibolie*, Kant is critical of space constructions such as Christian Wolff's physical monadology, where space results from a dynamical interaction, or diffusion of force, of spaceless, timeless monads and so-called primary, instantaneous forces. Kant calls this an attempt to "intellectualize sensibility" by imagining an intellectual construction of extension which itself assumes sensible intuitions. For example, assuming one tried to generate a dynamical construction of space from causal interactions of monads, it seems one would be bound to assume a causal interaction *between* the monads, so that the construction must fail, or drive one to assume the absurd doctrine of pre-established harmony.

Presumably Kant knew what he was talking about since he, too, had tried his hand at the physical monadology and the dynamical space construction in his Pre-Critical period and failed. More recently, it has been proposed that *all* dynamical, causal derivations of space and time (for which see Robb 1913, Reichenbach 1927/1956, Finkelstein 1969, Winnie 1977, Penrose 2004) are doomed to fail, even in advanced physics and novel approaches such as causal set theory, since dynamics and causation always assume extended length, or extended causal processes and interactions (see Earman 1972, Hagar and Hemmo 2013). If this were true, Kant would be right to call extension an intuition we bring to all extended representations of the world, rather than a concept, for apparently we have no conceptual grasp of it at all in mathematics, and we cannot understand how such a conceptual construction is possible without begging the question.

So this is still the basic challenge to philosophers: give a conceptual construction of extension that is consistent and non-question-begging and interpret this construction physically so that actual physical space-time can be seen to be a result of such a construction. Russell of course made a start on this project, and there is even a passage in the *Analysis of Matter* where he does consider getting rid of the embedding space-time coordinate system in relativity (Russell 1927/1954, pp. 67–69).

Evidence for a Leibniz-inspired constructivist program

A philosophico-scientific program for a construction of extension certainly existed in Germany, at least up until the time of Riemann and Grassmann. The construction originally took the form of what was called a "physical monadology," inspired by Leibniz's writings and pursued by Christian Wolff, Roger Boscovich, and the Pre-Critical Kant among others. Herbart developed a physical monadology in his *Allgemeine Metaphysik* of 1828–1829, which involved unextended point-like *Wesen* (beings, entities) and instantaneous forces, and where extension is traced out by an associative–dissociative relation in time, but which does not assume temporal extension, only present states and other adjoined present states serving as "images" of past states. Herbart is a direct historical-conceptual link to Bernhard Riemann (Scholz 1982) and is even mentioned by name alongside Gauss in Riemann's seminal 1854 *Probevorlesung* on the foundations of geometry. There is also some new evidence uncovered by Petsche (2012) for an indirect influence of Herbart on Grassmann through family friend Carl Scheibert. In writing his 1847 prize *Essay on the Geometric Characteristic*, Grassmann, however, looked back to Leibniz and claimed that the geometric algebra developed in his groundbreaking 1844 *Ausdehnungslehre* was Leibniz's "geometric characteristic," adumbrated in a famous letter to Huygens in 1679 (see Leibniz 1989). Grassmann's prize essay was written in response to a challenge to complete Leibniz's project for the characteristic, but the connection was more than self-serving I think.[2] Grassmann also claimed that because his geometric algebra described abstract entities and relations at a level *prior* to extension, they could be used to conceptually *analyze* the concept of extension, a goal he attributed to Leibniz himself and which is mentioned in the original letter:

> Finally, at the end of Leibniz's presentation [the 1679 letter] is yet another remarkable point where he quite clearly expresses the applicability of this analysis to objects that are not of a spatial nature, but adds that it is not possible to give a clear concept of this in a few words. Now, in fact, as is demonstrated throughout Grassmann's *Ausdehnungslehre*, all concepts and laws of the new analysis can be developed completely independently of spatial intuitions, since they can be tied to the abstract concept of a continuous transformation; and, once one has grasped this idea of a pure, conceptually interpreted continuous transformation, it is easy to see that the laws developed in this essay are also capable of this interpretation, stripped of spatial intuitions. (Grassmann 1844/1995, p. 384)

As we shall see below, this central motivation to develop extension "from scratch" without assuming an extended background is common to all of the thinkers in the constructivist tradition. A rough definition of constructivism might run:

---

[2] According to De Risi (2007) Grassmann algebra should be seen as the more direct descendent of Leibniz's characteristic in the 1679 Huygens letter and the late, c.1714, *Initia Rerum Mathematicarum Metaphysica*. Grassmann himself certainly believed he had perfected the geometric characteristic along the scientific and philosophical lines shown by Leibniz.

1. Geometric space is an analytical construction from concepts, *not* a basic intuition we must take for granted.
2. The structure of physical space and any rules for constructing it are empirical, *not* a priori matters.
3. A construction of geometric space must not assume the intuition of extension either in space or time, or in the dynamics of the construction, say by dynamical space-tracing processes.

J. F. Herbart: an important transitional figure

J. F. Herbart (1776–1841) was Professor at Königsberg and Göttingen. He was famous for his mathematical psychology of psychic forces and for his philosophy of spatial representation as well as his works on pedagogy (see Banks 2003, ch. 3). His central philosophical works were the *Psychologie als Wissenschaft* of 1824–1825 and the *Allgemeine Metaphysik* of 1828–1829. Herbart was a realistic, scientifically minded philosopher who advocated constructivist methods and who challenged the Kantian a priori drafting-board intuition of space and time extension. In the history of philosophy, Herbart is often considered a neo-Leibnizian because he worked on constructing space in a physical monadology, like Christian Wolff and the young Kant before him. For Herbart's place in the overall history of space perception in psychology in comparison with other figures, see Boring 1950, Hatfield 1991, and Banks 2003, chs. 3 and 4. In this chapter, I will focus on the particular Leibnizian line of influence I have identified (Banks 2013c), but the reader should be aware that there are other interesting constructive, empiricist views of space perception advanced during this time period, by Wundt, Lotze, and Helmholtz, for example the Lotze–Wundt theory of *Lokalzeichen* or local-signs (for which see Banks 2003, ch. 3).

In his *Metaphysik*, Herbart describes a community of point-like *Wesen* (things, entities) possessed of instantaneous forces as they push and resist each other. He then describes combinatorial, symbolic associative–dissociative extension-tracing processes among the *Wesen* that amount to a construction of extension, at first by building up a simple discrete extended series of dots called a Starre (rigid) line (Herbart 1964, §245) and moving on to three-dimensional *Körperlicher Raum* and the construction of continua.

Herbart believed, with Leibniz, that the "primary" forces of nature, at the metaphysical level of a community of *Wesen* beneath extension, were instantaneous or momentary and did not require an extended representation. *We* interpret the combinatorial relations of forces as the serial tracing of a line but the processes themselves are prior to any such extended interpretation. In the underlying community we find only an instantaneous array of primary forces active at any given moment.

How do we understand what he means by instantaneous forces beneath extension? If we simply take the notion of "force" straight out of mechanics, a freely acting force always requires some space to act through (producing energy) and some time to act over (producing momentum) and manifests in accelerations over extended space and time. Even the calculus (which Herbart knew well) will not help, since differentials are taken as limits of extended quantities and never quite cross over into the realm of the unextended or instantaneous. But here we might take a suggestion from the natural philosopher and metaphysician Roger Boscovich, who pointed out that we can imagine that there *are* no free forces at the primary level, but instead that each primary "active" force always acts against a countervailing "passive" force (1763/1966), freely accelerated bodies still resisted by inertial forces, perhaps, so they cannot be accelerated any faster. Boscovich proposes that the greatest active force is equilibrated by the smallest passive force in a small enough "snapshot" of time and space. Actually there are two possibilities: either (1) the stronger force will always overcome the weaker continuously and there never is a minimum, in which case we always have to work with extended differentials of force, or else (2) there are discrete quanta of energy-momentum, minimum block-like lengths and durations, which *cannot* be made smaller, even when the greatest force acts against the weakest. At that level all is in stasis: we can only assemble the quanta of extension out of static, combinatorial patterns of primary forces in equilibrium moment by individual moment, like stringing together the frames of a film where only the actual present moment is real at any given time. We are not assuming a minimum extended length, or duration, in which dynamical processes still go on, we have actually analyzed the minimum extension *as* a dynamical equilibrium point explained by the behavior of active and countervailing forces.

So let us suppose (2) to hold and this balance point among the primary forces to have been reached, and consider a snapshot of the community of *Wesen* at a given instant. Herbart sees two kinds of instantaneous relations for the primary forces: they either depend on each another (*Zusammen*), or they do not (*Nicht-Zusammen*). These are instantaneous causal relations of dependence or independence within each static snapshot. It is as if we can tell already in the static case whether a force causally abuts upon another opposing force or whether it simply acts as if the other is not there, independently. Herbartian psychology, for example, is based upon the notion of a simple "apperceptive mass" of psychical forces which are all united at a simple point strictly through these sorts of unextended and instantaneous causal relations and it seems likely that he took the notion over into his metaphysics as well when thinking of the unextended community of *Wesen*. Hence when these forces are put into a series of static combinatorial patterns (the rows in Figure 6.1) their instantaneous relationships, which we express in symbols, can bear the intuitive interpretation of the serial tracing of an extension. We need not, and should

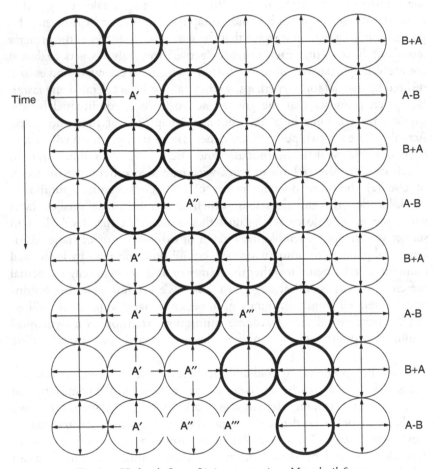

Fig 6.1   Herbart's Starre Linie construction, *Metaphysik* §245

not, think of these operations as involving motion, or dependence in time and space, but rather through abstract concepts of constancy and variation, as might be true of two functions (f, g): f is capable of variation along g, while g holds constant, and vice versa, generating an abstract "product space" f × g.

In his *Allgemeine Metaphysik* (Herbart 1964, §245), my reconstruction of which is shown in Figure 6.1 (Banks 2003, 2005, 2013c), Herbart imagines an "inchworm" process, tracing out a line of discrete points and dissociating gaps. Let **A** and **B** be the names of the present nuclei that do the tracing. (A + B) represents the operation of associating B to a fixed A, (B + A) the opposite, (A − B) represents "taking B away" from fixed A, dissociating B from A, (B − A) the opposite. This gives us one infinite symbolic sequence:

$$(B + A) + (A - B) + (B + A) + (A - B) + (\mathbf{B} + \mathbf{A}) \dots$$

We also need an operation to represent what Herbart calls the generation of *images*. Sticking with the intuitive interpretation, when the tracing process moves on we need a way to represent its "past" stages. That is, we need some other present nucleus to serve as a record or copy of where A has previously been, called A', A'', A''', and so forth. The copies are generated, in sequence, by the number of *times* we dissociate B from A and reassociate A with B. Finally, we can express the order and structure of the series of associations and dissociations by nesting the previous stages inside the later stages in one set:

$$(((((((\dots A''' + B''') - B''') + A'') - B'') + A') - B') + \mathbf{A}) - \mathbf{B}) \dots$$

I use + and − to represent more abstract symbolic relations of association and independence, or dissociation; it seems to me that we can also use the nested bracketing generated by the process to indicate a closer association between two points or images (A, B) and also to indicate that they are less closely associated with a third thing C: ((A, B), C).[3]

Herbart's generation of a row of dots is an attempt to analyze the intuitive content of extension through abstract concepts or even symbolic

---

[3] Following on a suggestion in my 2008, one might consider interpreting the bracketing as association and dissociation in making the transition from a discrete extension to a continuous extension, where we would somehow have to reproduce the cardinality of the continuum by taking the power set of all the entities in the set of discrete points. The power set is the set of all subsets, but it could also be interpreted here as the set of all combinatorial associations and dissociations of a set of elements like (A, B): ø (or no association), (A, B) both associated, (A), (B) both dissociated from each other, and so forth.

operations. Strikingly, so far as I know, he is the first to recognize in extension the need for an alternation between a *dissociating* operation, where his instantaneous forces are independent of each other, and association, where the forces are dependent. Extension, as in the gaps between Herbart's *Wesen*, is really just a kind of causal independence or dissociation of the primary forces from each other, which he calls a *Nicht-Zusammen*, where each is capable of acting independently of the other as if it were not there.

Herbart insists, over and over again, to the point of being maudlin, that his abstract symbolic processes can adequately represent serially generated spaces, but they need not be represented that way. This must have been a common objection to his construction, which he strains to answer in many places in his work. The symbolic processes are interpretable as the serial generation of points, like a tracing, but the symbolism also bears *all* possible interpretations and not just as the intuitive serial process. The familiar objection that these processes are *already* extended is thus avoided by pointing to the "abstractness" of symbolic operations in mathematics, as we emphasized above, or to the static patterns of forces that actually analyze units of extended magnitude into dynamical snapshots in equilibrium. By laying down symbolic or combinatorial operations instead of geometric ones, one does not commit to any intuitive interpretation of the abstract symbols and combinations, and this is one kind of answer to the charge of circularity.

Herbart's "inchworm" process strikes me as a primitive sort of symbolic program, automating a series of steps and operations, but which is also present as a *complete object*, indifferent to serial order of any kind in the whole set of nested stages and images. Any stage of the process could be represented as the present one, the centered, present perspective of a particular track through space, by altering the order of nesting to make it the new center with the other stages or images nested around it. The process can also be reversed in direction by pushing through a minus sign, giving us a row of $-A$'s, or, in keeping with our conventions a row of B's, generated in the reverse order from A (suggesting that $A = -B$ and $A + B = 0$). The operative assumptions seem to be these:

1. associative and dissociative operations in an alternating pattern;
2. a copy operation which allows other present stages to serve as nested records;
3. universality: any stage can be considered the present stage of nesting, by rearranging brackets; any image can be made the present relative to

other stages by reassigning the primes; any point can be the present center of a particular track through the manifold;

4. absoluteness and relationism: The forces and their combinatorial relations can be thought of like vectors are today. All of the constructions will be directional- and basis-independent and will not depend on the choice of coordinate systems, or reference frames from which the process is viewed, in either space or time.

Herbart also provided an analysis of extension in psychology, for extended sensory manifolds of sight, color, sound, and touch, which seems to have been an important influence on Riemann in developing a notion of a manifold *prior* to the introduction of a metric (see Scholz 1982). These manifolds are built up point by point through processes Herbart called "reproduction series" (Boring 1950, Banks 2005). For example, Herbart imagined an individual color point changing in hue as it "moves" around the color manifold on its own track. The previous stages of the process are reproduced in memory and the memory images are added, in an associated–dissociated combinatorial pattern also preserved in memory images, to the present stage of the point. Every such path traced in the color manifold thus makes up an "individualized perspective" in which colored points are generated in a series, from within the manifold. There is no outside view of the color manifold as a solid with all its points filled in; the manifold *is* just the sum total made up of perspectival views from within, along all these separate individual tracks, without any "master view" from the outside.

As a psychologist, Herbart realized he also had to explain the origins of spatial representation. How are the purely symbolic, combinatorial processes of intelligible space in the *Metaphysik* visualized as spatial manifolds in the *Psychologie*? The changes of a colored point, or the tracing out of a row of dots in the *Metaphysik*, do indeed make up an extension of a primitive sort, but as Herbart himself emphasizes, this extension is *intuited* only in the memory or records of non-present images. The present stage is always an unextended moment or a point, a momentary arrangement of primary forces in their instantaneous relations of dependence and independence. Herbart claimed that an actual intuitable extension of a length required *two* reproduction series, crossed over each other. When two reproduction series are crossed over, or associated and dissociated with each other, this more complexly structured process appears in an intuitable way, from inside the manifold, as a "length" made up of separated points.

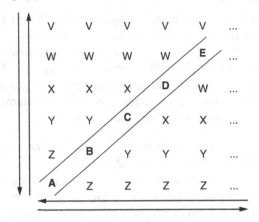

Fig 6.2   Herbart's psychological reproduction series

I call this the "wire argument" for short. Consider a wire seen head-on as a point. The extension of the wire is not intuitable unless the wire is turned through some independent direction, associated with the first, but also allowing us to *dissociate* the points on the wire from each other. Two reproduction series do that but in a causal sense. From within the manifold, running along a track, one traces out and reproduces the points visited not in one direction but in two. In Figure 6.2, imagine one series A, B, C, D, E with each letter also potentially associable with any of the range of letters run through in the column V, W, X, Y, Z. As we pass from A to B, we can imagine traversing the whole range of the other series, setting them apart from each other. Nevertheless, the closeness and the order of the association between A and B are still preserved, since no particular causal associations are created between A and any individual letter of the other series. For more on this and Hermann Lotze's contributions to these sorts of constructions in psychology see Banks 2003, ch. 3).

The difficulty in seeing such an analytical construction as actually extended in intuition was remarked upon by psychologists, including William James, and by Gary Hatfield in his comprehensive study of spatial representation in psychology (1991). Again, I think this was because Herbart's readers expected to stand *outside* the whole space and observe its extension from the outside, like a box, whereas Herbart had always insisted that the extension of space only appears from *within*, where the point of observation is an internal one within space from an origin of a perspective to other points encountered on other tracks. Mathematicians sometimes call this an "intrinsic" view of space, thinking of Gauss, who developed a theory of curvature that lies within the surface studied, instead of the ambient space. This desire to "see in from outside" harks back to the science museum fallacy, say, of representing the *entire* universe, while we seemingly stand outside in nowhere. From Herbart's point of view, *all* realistic spatial representation is intrinsic. We have no idea what space would *look like* from outside space and it is a metaphysical error to think we could. A "space" is just the sum total of all the perspectival tracks or tracing processes active inside it, not a finished extended object or box seen from the outside.

As different as Kant's view of space is from Herbart's, it is clear that both philosophers interpreted space and time extension as primarily a form of *representation*. Space and time are a perspectival system of objects-to-a-subject, in which a subject traveling along his own track represents perspectival objects or spaces from various points of view within the same space in which he encounters the objects. This underlying "subject–object" structure seems to be intellectually fundamental to the representation of both space and time and of a piece with the intrinsic view of space, i.e., that spatial representation only makes sense by taking up a series of vantage points from *within* space, in which other spatial objects and regions are encountered, not by viewing it as a finished box from without. Space for Herbart is also dynamic: its instantaneous state is always being constructed, moment-by-moment, by the processes that trace it out. There is never anything like a "finished manifold" of points one can just take off a shelf. I think this may be a crucial difference between these philosophical constructions of space and time and the way some contemporary mathematicians view manifolds as all given in advance, or embedded in some ambient space for study, without really taking the intrinsic point of view.

Riemann on multiply extended magnitude

Bernhard Riemann discusses the problem of what he calls "multiply extended magnitude" in his famous 1854 *Probevorlesung* lecture "On the Hypotheses that Lie at the Foundation of Geometry" (Riemann 1867). He credits Herbart by name, alongside Gauss no less, and mentions some "philosophical investigations" that helped him. Most scholars now think that Herbart's psychology of sense manifolds of color and sound and reproduction series had a greater impact on Riemann than the construction of points in the metaphysics with which Riemann said he disagreed but did not specify in what respect; for example, he may simply have objected to Herbart's fallacious proof that space had to be three dimensional (Russell 1897/1996, Scholz 1982). Riemann explicitly takes color space and (in notes for the lecture) the tone row (see Scholz 1982, Banks 2005) as illustrations of three- and one-dimensional manifolds respectively.

Riemann also says right away that he does *not* intend to assume the intuitive drafting board of space on which to do constructions, but that he intends to derive the special case of space as a concept of multiply extended magnitude from "general concepts of quantity" (p. 255). This language is very nearly a declaration of war in the Kantian climate of the time, though perhaps not in Göttingen, where Herbart had taught. The first task here is to define a manifold extension from concepts. The second task is to give definitions of intrinsic curvature and measure determined from within this extension, by introducing a metric as a function of the coordinates in geometry, or rigid meter-sticks or light beams as standards of measure for physical space, as Riemann is aware.

The first task is the one which interests us here, since it concerns the very origins of extension. A manifold, Riemann says, consists of various "modes of determination" (*Bestimmungsweisen*), a minimum of two. The discrete points, or continuous elements, of the manifold are determined by transitions, discrete or continuous, between these modes.

> Notions of quantity are possible only where there exists already a general concept which allows of various modes of determination. According as there is or is not found among these modes of determination a continuous transition from one to the other, they form a continuous or a discrete manifold; the individual modes are called in the first case points, in the second elements of the manifold. (1867, p. 255)

In the color manifold, the modes of determination are qualities, colors. In the tone manifold the modes are pitches of high or low. In a manifold like a line the modes are directions like right and left in a one-dimensional continuum. So, for example, as the colored point changes, its mixture of color qualities or directions exchange continuously and different colored points are thereby determined. As the tone slides from high to low along the tone row, it exchanges directions of high and low, like directions of forward and backward on a line. What determines the basic structure of a manifold for Riemann is the number of these possible qualities or directions of the manifold, and the properties of the transitions, which determine whether points or elements will be generated by the construction. For example, the two-dimensional line depends on its having two directions, forward and back, and a continuous transition between the directions:

In a concept whose various modes of determination form a continuous manifold, if one passes in a definite way from one mode of determination to another, the modes of determination traversed constitute a simply extended manifold and its essential mark is this, that in it progress is possible from any point only in two directions forward or backward. (Riemann 1867, p. 257)

Riemann also indicates that predicates or class concepts can serve as "modes of determination" in the most general manifolds by which individual objects are classed together, identified, ordered in series, and finally counted. Why must we have at least *two* modes of determination? Probably because individuals that all fall under one predicate or class concept cannot be differentiated further. We want to be able to further differentiate individuals that are all alike in some property. This we can do by introducing some other predicate to set them apart. All men of a certain height can be separated by weight. A third predicate, wealth, can separate them still further.

Riemann thus generalized the idea of a manifold as far as it would go into the abstract realm of concepts, both quantitative and even qualitative concepts, like colors, or just predicates. You can have manifolds of directions, qualities, or even physical properties, the structure of the manifolds defined by their number of modes, or dimension, and the nature of the transition, discrete or continuous. These properties can be independent, which is required whenever we want to differentiate an individual from another. But the properties are clearly also dependent at points or individuals determined by the transitions. It would thus seem that the modes are capable of dependence in individuals, and independence in the interstices, and that the transition combines both the dependence and independence of the modes, alternating as they transition from one individual to another.[4]

---

[4] With qualities it is easier to see how they could be sometimes dependent and sometimes independent in the required way, but on the line we have only opposed directions that are dependent everywhere. I have thought about this problem in Riemann for a long time and come up with the following explanation. When the line is presented as a one-dimensional continuum of points or elements, we are attending to the points or individuals but we are not thinking of the underlying modes that generated them. Hence, focusing on the individuals generated, it seems to us that these modes must be dependent everywhere, since we only see them where they are dependent at points or individuals. But if we attended more to the extension *between* the points or elements, we would see the need for another series of dissociations to set the points or elements apart from one another, and see that actually the modes must also be capable of dissociation in the interstices. As counterintuitive as it sounds, even directions like right and left must somehow be capable of dissociating from each other (as I discussed in Banks 2003, 2008).

This is why I believe Herbart's constructions, whether in the *Psychologie* or the *Metaphysik*, *must* have influenced Riemann somehow, because they seem to operate in the same conceptual universe, involving some sort of alternating, combinatorial associative–dissociative relation to generate extension, and they both see a conceptual construction of extension as primary to the study of space in geometry or physics, which lends an important philosophical dimension to Riemann's work in addition to his mathematical innovations.

Turning to the question of measure, Riemann says that in the case of a discrete manifold, the definition of measure is welded into the manifold extension itself. We have only to count up the points traversed as the modes switch discretely in their values. For a continuous manifold, Riemann says that measure can only be defined by selecting a standard of measure external to the basic manifold of points or elements, like a metric defined over the coordinates, or meter-sticks and light rays for physics. He restricted this condition by demanding that the space exhibit *constant* curvature, positive, negative, or zero, a requirement we do not retain today as we accept spaces of non-constant curvature caused by matter and energy in the theory of relativity. However, Riemann did show how the metric element of Euclidean space, the straight line, was only a special case of a metric element which need not be straight: it could be a curved-arc element instead, for example, thus contradicting Kant's claim that the straight line is, a priori, the shortest distance between two points. And Riemann also hinted that future measures of space might simply be discovered empirically within the manifold itself by the natural forces, a view he attempted to defend in his fragment on natural philosophy, part of his *Nachlass*, and which many consider an early anticipation of Einstein's general theory.

Hermann Grassmann's *Ausdehnungslehre*: overview

Many of the constructive ideas above find a natural home in Hermann Grassmann's masterpiece, the *Ausdehnungslehre* of 1844, 1861 (Grassmann 1884/1995; for illuminating studies of the system see Crowe 1967, Zaddach 1994, Schubring 1996, Swimmer 1996, Browne 2009, Petsche 2011). Grassmann's work is devoted to an analysis of the concept of extension in all generality, of which space forms only a special case. Thus he explicitly breaks Kant's stricture, as Riemann did, by "subordering" space under the more general concept of extension, bringing space under concepts after all. This bold move was not lost on his Kantian critics, who

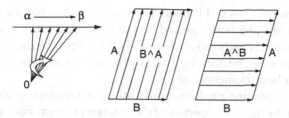

Fig 6.3   A bound vector and a free bivector area

reacted swiftly. For example Ernst Friedrich Apelt defended the party line in a letter to Möbius of 3 September 1845:

> Have you read Grassmann's strange *Ausdehnungslehre*? It seems to me to be based upon a false philosophy of mathematics. Such an abstract theory of extension which he seeks can only be developed from concepts. But the source of mathematical knowledge lies not in concepts but in intuitions. (Petsche 2012, p. 187)

Grassmann's chief innovation is the exterior product (∧), an anti-commutative extension-building operation, which takes the geometric product of two magnitudes.[5] Starting with the point, we can build up higher and higher extensions without limit. Points α, β, for example, multiply to a directed line segment, α ∧ β, which we call a (bound) vector, and (free) vectors A, B multiply to a parallelogram area element, A ∧ B, known as a free bivector (see Figure 6.3). This operation can be continued to trivectors and n-grade entities and thus we are not limited to any number of dimensions by intuition or our ability to visualize space psychologically.

Grassmann himself separated the algebra of points and bound entities based upon points from the algebra of free vectors and entities based on them. Some authors continue this tradition by separating the projective space of points from the affine space of vectors, both metric-free. Other authors,

---

[5] The formal properties of the exterior product are defined as follows (following Browne 2009). For elements of grades m, k, r:
  1. Associativity: $(\alpha_m \wedge \beta_k) \wedge \gamma_r = \alpha_m \wedge (\beta_k \wedge \gamma_r)$
  2. Identity: $\alpha_m \wedge 1 = \alpha_m$
  3. Anticommutivity: $\alpha_m \wedge \beta_k = (-1)^{mk} \beta_k \wedge \alpha_m$ (anti-commutative if both elements are of odd grade)
  4. Distributivity over addition: $(\alpha_m + \beta_k) \wedge \gamma_r = \alpha_m \wedge \gamma_r + \beta_k \wedge \gamma_r$
  5. Scalar multiplication: $(a\, \alpha_m) \wedge \beta_k = \alpha_m \wedge (a\, \beta_k) = a\, (\alpha_m \wedge \beta_k)$

such as Swimmer 1996 and Browne 2009, emphasize a combined interpretation which includes both free and bound entities within one algebra. For example, a simple combined algebra can be built up from the basis of a zero point and three free vectors, by exterior-multiplying all with all to generate higher-order basis elements.

Another important mathematical feature of Grassmann algebra is its duality, also due to the properties of the exterior product. For each element there is a "dual element" above, which exterior-multiplies to the whole dimension of the space. There is even an extension-lowering operation dual to the exterior product, called the regressive product (v), which intuitively takes the "intersection" of higher-order elements, bringing them to lower order. Elements of levels 4 and 0, 3 and 1, and the free and bound elements of level 2 are all duals since they multiply pairwise to the dimension of the space, 4. In treatments which follow Weyl (Zaddach 1994, p. 100), the whole algebra has a parallel dual algebra, for example of vectors and differential forms dual to vectors, defined in such a way that corresponding elements from either algebra multiply to a real scalar number. This is in fact the way the familiar inner product, and included scalar product, of vectors are truly defined. I took a different approach by allowing the volume element to determine the measure (Banks 2013c) and then just took the dual to get real numbers. This helpful property of duality is fundamental for the definition of an inner product and introduction of a metric based on the inner product of the basis elements with each other which can assign a measure to each element of the space. The Grassmann algebra is so fundamental that the inner product can actually be defined explicitly by exploiting the duality of the algebra (see Browne 2009). When the metric is defined, the elements get their measure based on a determinant, a general measure of volume which can be extended to entities of any dimension, vectors, bivectors, trivectors, what have you (ibid.). As I have shown, a determinant even "works" for unidirectional Riemannian "modes of determination" prior to points and produces a bidirectional length between two endpoints indifferent to direction, shown in Figure 6.4 (Banks 2008):

The geometric entities, and the basic set of algebraic operations on them, like sums and differences and the exterior and regressive products, can all be defined without the assumption of a coordinate system or a metric, and are all extremely basic, tensorial entities, not only independent of the coordinate system but even conceptually *prior* to it. They are well suited for interpretation as physically real entities, like forces, but also as

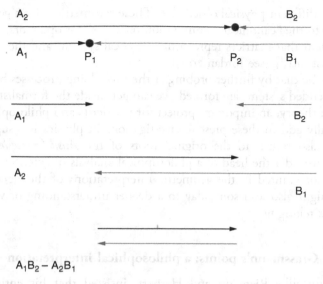

Fig 6.4   The determinant for Riemannian modes of determination (Banks 2008)

general slots one can fill with a variety of entities: vectors, differential forms, operators, even transformations. When metric notions are introduced in Grassmann algebra, and the orthogonal complement and inner product are defined, the powerful Clifford algebra of rotations can be introduced, which incorporates Hamilton's quaternions, the complex numbers, and the algebra of matrices (see Hestenes 1999, Doran and Lasenby 2007, Browne 2009).

These ideas are currently being applied to give an intuitive geometric interpretation of quantum mechanics. For example, Hestenes recommends interpreting the iℏ of quantum theory as an oriented area which harkens back to the origin of the "quantum condition" as an area in the phase plane of a physical system. A physical system, such as a resonating atom or an harmonic oscillator, can be represented by a matrix of its various eigenvectors and eigenvalues, which is like an exterior product of its column vectors, or its row vectors, in a given basis. As has been known for a long time, the same matrices that are used for representing physical properties of the system, such as energy, momentum, spin, position, can also be used as operators to transform or rotate the system in various invariant ways inducing a change of time, position, direction, and velocity or a change of basis to a different physical observable. These symmetries give a geometric meaning to the "canonical commutation relations" for operators, a crucial requirement that matrices representing physical systems and their observables must satisfy (see Jordan 2005).

It may be that by further probing at the underlying processes by which these extended systems are formed, we can get inside the formalism of the quantum theory, an important project for physicists and philosophers. All I can really add to these present investigations in physics is a suggestion that we also return to the original roots of the *Ausdehnungslehre* and consider it under the head of a philosophical analysis of extension, which we take for granted in the geometric interpretations of the algebra, but which might also lead some day to a deeper understanding of what the physics is telling us.

Grassmann's points: a philosophical interpretation

Grassmann, like Riemann and Herbart, insisted that his entities and products *not* be understood primarily as extended geometric entities in a background drafting board of space:

> It is important to keep in mind that all of the elements so generated are not to be conceived as already given otherwise, as perhaps in the theory of space all points are originally given through the presupposed space, but rather as being generated from scratch. (Grassmann 1844/1995, p. 50)

In fact Grassmann deliberately used an abstract philosophical language of "systems," "evolutions," and abstract "constancy" and "variation" to avoid assuming extended intuition. This language severely hampered the reception of the work by mathematicians, even by Hamilton, who was philosophically inclined (Crowe 1967). Grassmann could have left out the philosophy and stuck to the math, but it was clearly essential to what he thought he had achieved. We know he was a serious student of philosophy (Lewis 1977) and he may well have known Herbart's work, perhaps through Scheibert (Petsche 2011). Grassmann algebra is used today like any piece of mathematics and nearly always takes an extended interpretation. The processes of tracing extension are never seen the way they were originally intended, as the actual tracing out of extended magnitudes from scratch.

What we will want to do now is apply Grassmann algebra to our problem of constructing extensions from elementary event particulars or point-events. Let's begin at the simplest level with Grassmann's algebra

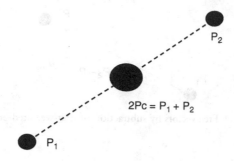

Fig 6.5   Summation of two equal weighted points

of points and free vectors.[6] A point is defined by an arbitrary zero point and a free vector which carries the origin to the new point from the zero, $\mathbf{P} = P_o + \mathbf{v}_P$. Points are bound to their positions and cannot move. They also have a weight, $a\mathbf{P}$, positive or negative, like charges, with 1 as the unit weight of a geometric point with position only, $1\mathbf{P} = \mathbf{P}$. Points sum barycentrically, to their center of gravity. Two positive, equally weighted points sum to a point of double the weight on the line between them (see Figure 6.5). Two *unequally* weighted points sum by their respective weight-distances from their center of gravity.

Two unit points can also be subtracted, $\mathbf{P}_2 - \mathbf{P}_1$, cancelling the origin and leaving a free vector from $\mathbf{P}_1$ to $\mathbf{P}_2$: $\mathbf{P}_2 - \mathbf{P}_1 = (P_o + \mathbf{v}_{P2}) - (P_o + \mathbf{v}_{P1}) = \mathbf{v}_{P2} - \mathbf{v}_{P1}$. This free vector is called a point at infinity and has only direction and no weight or location (see Figure 6.6).

The two most common physical interpretations of the point calculus are Archimedes' law of the lever and the determination of the center of mass of a system. But it is also possible to interpret the point as the potential energy (Banks 2013c). For example, imagine starting with a particle at some

---

[6] Following Swimmer (1996) we can characterize the points as follows. Let {A, B, C ...} be points, let $A = [A] + \mathbf{v}_A$, meaning let it have a weight [A] and a position vector $\mathbf{v}_A$. If B and A have equal weights, let B – A be the zero-weighted point, or free vector, $\mathbf{v}_B - \mathbf{v}_A = a(\mathbf{v}_B - \mathbf{v}_A)$. Let the free vectors of level 1 obey all of the usual rules of a vector space, addition, subtraction, multiplication by scalars, inverses, and zero vector. Let the points obey the following rules for barycentric addition, subtraction, multiplication by scalars and additive inverses:

1. $A + B = B + A$
2. $A + (B + C) = (A + B) + C$
3. $A + {-A} = 0$
4. $[\alpha A] = \alpha[A]$
5. $(\alpha + \beta) A = \alpha A + \beta A$
6. $\alpha(A + B) = \alpha A + \alpha B$
7. $\alpha(\beta A) = (\alpha\beta)A$

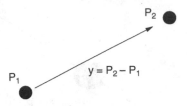

Fig 6.6   Free vectors by subtraction of unit-weighted points

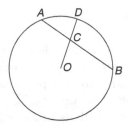

Fig 6.7   Grassmann's color manifold and mixing law

chosen origin and imagine the vector as a transition, doing work by taking the point a weighted distance from a force center, representing a potential jump from the center to the point. The point now has a potential energy relative to that center.

Grassmann gave an abstract interpretation of the point calculus, which he used for deriving his famous law of color mixtures (see his original diagram in Figure 6.7). This is a color-wheel theory, approximately true of real color vision, in which he represented white at the center (O) and represented the hue of each color (A,D,B) by its direction from O. The weight of the point represented the saturation, or mixture of white to the pure spectral hue at the rim.

Grassmann's law of color mixing states that the colors add barycentrically just like weighted points. This has the consequence that every color will have a complementary color and that a mixture (of lights not pigments) will always contain some added white (as A and B sum to C, not D). Clearly this is not an extended space filled in with colored points, for no extension has been defined. It is, however, an abstract space for the classification of points by their individual qualities, quite similar to the very general manifolds of qualities or properties introduced by Riemann.

My preferred philosophical interpretation of a primitive structure of points is a manifold of point-events with their individual "qualities"

marked by potential jumps. A quality thus becomes like the manifestation
of a force, and potentials become what we called above a power. The
quality marking an individual point-event is a potential jump from a zero,
chosen arbitrarily. But the same jump can be marked from all other zeroes,
and any point-event can be the origin of our perspective on the event. Each
individual point-event is representable by the sum of its individual qualities, or jumps it represents from all other point-events chosen as a possible
zero. We can think of the point-event as having individual qualities
"radiating" out of it like spokes, giving it a uniquely identifying signature
(see Figure 0.1).

The structure so far is a perspective space of individualized point-events
and perspectives in which we allow any point to be a center of perspectives.
Since each event is instantaneous, and no extensions are defined on it, the
manifold is not extended or filled in. It is more like an abstract space for
the individuation of point-events. This is a projective space, because there
is no parallelism of the directions or vectors that is invariant across
perspectives, nor is there a measure or metric property for distance or
weight of the point from the center. So if the various spokes represent
individual qualities of the event, the qualities will be different from each
perspective. The most we can say is that each quality we attribute to an
event is a jump viewed from a given zero somewhere else. That is like
saying the same power manifesting one way in one qualitative manifestation could have manifested differently from a different viewpoint. There
is little structure, but the multiplicity of all views, or directions alone,
suffices at least for a basic identification of all points. For example, point-
events that all lie in the same direction from one zero can be differentiated
by taking up other vantage points elsewhere (see Figure 6.8).

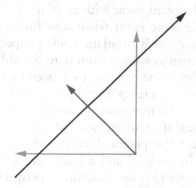

Fig 6.8   Differentiating point-events by their directions

In my view, the Grassmannian point-events are the best conceptual reconstruction for what Mach and Russell meant by "elements" or "event particulars." I see a very clear connection between the characterization of point-events through potential jumps or forces and Mach's notion of a physical element cum force cum potential change, and also I see the qualities of the point-event as representing real forces manifested in the jumps, similar to the way Russell thought of the occurrent qualities of event particulars as manifestations of effects or interactions. The summation of weighted points barycentrically makes abundant sense if the summation of points is really to be understood as the summation of their "effects" to their combined effect, or alternatively as the decomposition of a combined effect into independent events, where the order of summing of effects does not matter.

Of course, many philosophers and scientists would insist that events, powers, and manifestations are not what is primitive; rather, there are particles and fields which possess the powers that lead to the manifestation of those powers in interaction events. They would say that the *particle* jumps in a potential field and this is what the language of potential jumps and forces really expresses. The realistic empiricist view is of course the reverse: the particle is rather a kind of event, i.e., of measuring a certain quality of a particle-event from a certain perspective. So far as I know there is nothing in the actual physics to prevent this kind of event-centric rather than particle-centric interpretation.

As long as we are only interested in the identification of point-events by their jumps or qualities from all perspectives our manifold remains primitive. We can now go further and establish the perspective invariance of these point-events by exploiting their "vectorial" nature. If we introduce the idea that the point-event is something objective and perspective-independent, then the point-event with all of its qualities seen from one perspective is still the same event when seen from another perspective. Something about the event is invariant across perspectives. Say an event occurs and is represented as a jump from o_1 to B and has a certain weight or potential energy determined relative to B. Seen from A, the weight will be different, since now it is a jump from a point A to o_1 and then to a point B, where B is also seen as a jump from A. But if we rescale those A-centric events from B's point of view, we will get the same values as before. Given a certain zero, the potential energy of the jump represents a combined quantity (its weight) and a level value (vector) for the jump from the zero. These weights are determined perspectivally from zeroes, so they will change based on who observes them from where and so will their qualities. But we can demand that the same event be expressed in many different equivalent ways for other observers from all possible perspectives. Starting with the idea that the potential energy of a point-event is something objective and perspective-independent, we define a series of equivalences on the primitive perspective space. Then, just as a vector can be decomposed in any basis, we demand that a point-event be expressible as an event in all perspectives, all points of view on the event from all other zeros.

As real happenings, point-events do not depend on any particular point of view. They are also perspective-independent, since we can re-express them from a variety of different viewpoints, the same way we can re-express a vector in different bases. If we do not specify a perspective for a point-event, we get a kind of swarm of equivalent events in all possible perspectives, and points expressed as all possible sums and decompositions, which all coalesce into one when a particular causal perspective is chosen. The other possible perspectives are merely possible ways to understand the event from other objectively valid points of view as if they had happened that way. These other events and their qualities don't actually happen, but events are objective so we can consider all possible ways the given event *could* have happened, had the viewpoint or circumstances been different. In that sense, a broad notion of physical objectivity (that it doesn't matter from where you observe the same event) is complemented by the idea that only particular events and manifestations are concretely real for any given perspective.

Before we leave this perspective space of events and their manifested qualities, I wish to point out once more the similarity between this structure and what Kant claimed was the basic intellectual "hard core" beneath all spatial and temporal representations (discussed in relation to James in Chapter 3 above). The point-events in their perspectival relations do indeed exhibit that basic "subject–object" structure Kant considered fundamental for the representation of any objective reality: every point-event is a possible center of perspectives or a zero for representing all other point-events within the same space; each time instant or happening is a point of view from which to represent all other time instants or happenings. These minimal, intellectual conditions do not require an extended representation of space, but are purely intellectual conditions for a structure like the point-manifold, or other abstract spaces of classification of individuals, like Riemann's, to satisfy in general as a representation of nature, even more general than space and time.

Raising extension: the exterior product

Grassmann's achievement, I believe, was to explain how to raise an extension from this kind of very abstract unextended manifold of points in quality space to a real extended space of measurable, extended invariants, such as extended physical systems and their measurable properties. We begin in the point-manifold above, where point-events are simply classified by their qualities and perspectives, gradually adding structure, as we make the stronger assumptions of affine and then metric space (see Banks 2013c), and we end with a real extended space of physical systems and observables, all by using the exterior product operation over and over again to define systems of gradually higher order, beginning with unextended point-events.

The key to the construction is the sheer primitiveness of the exterior product, which, like its dual the regressive product, requires neither coordinate systems nor a metric for its definition. All we need to assume is the parallelism of directions, or an affine space, so that a vector always remains parallel to itself through affine transformations, as it does not in changes of position in perspective space. Imagine for example taking a grid of squares and squashing it. Although the squares become diamonds, and the angles are not preserved, two parallel vectors in one system remain parallel to each other in the other. And of course, the usual vector-space properties of identity, inverses, sums and differences, and multiplication by scalars apply. The structure is an affine space, in which the concept of position-independent *direction* makes sense but not a distance or a weighted point. With these assumptions in place, the exterior product can operate to create higher-order extensions even before we have definitions of orthogonality, inner products, and metric, showing that extension is actually prior to all of those concepts.

How does the exterior product actually work? Referring to Figure 6.3, in the event interpretation, we hold the first point α fixed and change its other qualities through a range, or spectrum, until the event qualitatively matches the individual endpoint β. This qualitative change of state has the property of being extended because we associate all the individual points through their common quality or mode along line $\alpha\beta$, and *dissociate* them through the range of other qualities passed through in a spectrum. We then reproduce and "add up" all of the intermediary points produced by this associative–dissociative process and obtain the extension $\alpha \wedge \beta$. What we have is a simple process of change between the points represented all together as a single process or what Grassmann calls a "system," a bound vector. This is a basis-independent object that still holds even when we rotate the whole extension into another perspective and express the exchanging qualities in a different basis of other qualities from another zero.

In the second case pictured beside it in Figure 6.3, we multiply two "free" vectors or directions by holding one free vector fixed and sliding the other along parallel to it. Free vectors are obtained by subtracting points and are simply directions or directional qualities, like Riemann's modes of determination. The only requirement is that they be capable of variation independently of each other. We cannot exterior-multiply dependent directions which "point" in the same direction ($A \wedge A = B \wedge B = 0$ is a defining property of the exterior product). In the multiplication of the two directions, one direction is thought of as holding constant through the change of the other. Grassmann's original gloss on the exterior product of vectors (1844/1995, pp. 50–51) describes how the whole range or "system" of determinations made by A is changed through the whole range of determinations of B, so that a range with A fixed is associated with each individual value B can take, setting the values of B apart by associating each of them with the whole range of A (see also 1844/1995, p. 171). This prevents the creation of any associations between any particular value of B and any particular value of A. This is built into the structure of the extension: every particular value of A can only be determined as associated with a *range* of the values of B, and vice versa. We obtain a different (and opposite) extension if we hold B's direction fixed and change A's instead, $AB = -BA$ (a non-commutative property which follows from $A \wedge A = B \wedge B = 0$ by multiplying $(A + B) \wedge (A + B) = 0$).

There are some caveats which apply if we are to think of these extensions as generated by underlying "tracing processes," since these processes are not uniquely determined by the extensions. As the algebra reveals in the case of the bivector parallelogram, the area is not of any determinate shape, since these are all equivalent: $A \wedge B = A \wedge (aA + B) = (A + bB) \wedge B$. The diamond-shaped area can be squashed or expanded in affine space without changing the exterior product. Also, the tracing process itself is indeterminate, since the same area element could have been traced out by another process with a change of basis, so long as it spans the same space. Bringing in the dual regressive product, the area element could equally well be expressed as the *intersection* of a trivector volume element and a planar element. Nevertheless, as Grassmann himself points out, the generating process, though indeterminate, does leave two determinate marks on the generated element, the extension and its handedness or direction. These marks suggest that some features of a tracing process are indirectly responsible for the extension even if we cannot say exactly which processes they were. Grassmann says "they [the produced extensions] appear as different since they are generated by different evolutions" (1844/1995, p. 50). This non-uniqueness also makes sense, however, in the same way the basis-independence of a vector makes sense. Just as the vector does not have a unique decomposition in any basis, so too the produced extensions here do not have any particular construction in a particular set of basis elements and operations, since one could always rotate or transform the generating process around and get a different one. So the generating process *must* be non-unique if it is also to be objective and basis-independent: the two features go together.

These generated extensions, starting with simple potential jumps, to processes like bound vectors, or area and other elements, gradually lead us via the exterior product to the construction of more and more complex entities, and finally to the matrices that represent physical objects and systems in extended space and time and their measurable observables in different bases. At each stage we consider (1) how the extension is produced from the basic "elements" (Mach), "event particulars" (Russell), or point-events; (2) the question of the extension's invariant, basis-independent properties or perspectives which allows us to represent the extension as something objective and perspective-independent; (3) the question of what extensions we can actually measure quantitatively in a given perspective.

Conceptually and historically, I believe it is impossible to ignore the similarity between Grassmann's exterior product, Herbart's construction of the Starre Linie and reproduction series and Riemann's transitioning modes of determination. The idea of a basic combinatorial construction of extension involving points, or even the simpler modes or qualities that determine points, seems to have led all three of these thinkers to some remarkably similar ideas. Oddly, so far as I know, Grassmann's system represents the last conceptual-philosophical efforts in this constructivist direction originally pioneered by Leibniz. Grassmann is ignored by Russell, although he was probably familiar with Grassmann's ideas, especially since his *Principia* collaborator A. N. Whitehead had written about the algebra earlier. There are of course many other efforts to construct space-time, such as those of Russell or Carnap, or the many contemporary constructions mentioned above, but none of them deal with the philosophical problem of extension in this head-on way that Leibniz, Herbart, Riemann, and Grassmann do. Unfortunately for philosophers, but fortunately for physics, Grassmann algebra was quickly retooled and adapted by

Gibbs and Heaviside into extended three-dimensional vector algebra and analysis, which became standard (see Crowe 1967). The vectors that represent physical quantities are dealt with as already extended things and their geometrical relations are explored through an algebra of extended operations, such as the algebra of rotations introduced by Hamilton and Clifford. The historico-conceptual philosophical tradition in which Grassmann's ideas were developed has now been completely expunged from the way they are understood today. This is usually what happens in the history of science, and this is often good, since naïve visualizations and models, along with motivating but misguided philosophical ideas, may add nothing to, or even detract from, the pure content of ideas. But sometimes this is not the case, if the philosophical ideas and the mathematical or scientific ideas are continuous with each other, and should *not* be cut away, as I suggest the case of Grassmann shows.

Extended representation: lessons from *Flatland*?

So *one* way to think of the construction of extension is the pure "Leibnizian" way, in which extension emerges from a metaphysical level beneath extension, consisting of instantaneous unextended point-events in a primitive manifold with only the barest intellectual structure of a system of perspectives. We then serially construct extensions in this underlying manifold, adding up the stages of the extension-tracing processes in memory, where only the present point-event or the instantaneous stage of the process is actually real. The rest is imagined or indirectly represented as a past trace of a present tracing process. The primitive manifold of point-events does not look anything like a space or time. It "looks like" a swarm of individual unextended events flashing in and out of existence, rather like the Grimaldi Rimini and Weber theory of quantum events (for which see Bell 1987). What we think of as a present trace of a "past event" is simply some other present event we press into service and call a trace, where even the relation between present events and past traces is a phenomenon of the present. The spatial representation of other events and other perspectives besides the one we presently occupy is indirect and depends upon the assumption that the same event here has an objective meaning from any other perspective somewhere else, even though events only happen in an individual concrete way and from a single perspective only.

When it is objected that we cannot actually imagine what this very abstract level beneath extension is like, the theorist (like Leibniz, or

Herbart, or Grassmann) can plead that what we represent to ourselves as a tracing process is only an intuitive representation of a deeper, more abstract "symbolic" order, of combinatorial "associations and dissociations," abstract "functional" constancy and variation, and other notions developed from concepts alone without assuming the imagery and intuition of extension. This is Grassmann's approach and explains his use of abstract language and substitution of symbolic relations for intuitive geometric ones. (I also followed this approach in my 2013c paper.) Notice that the extension theorist cannot actually explain, or imagine, what is going on at that level of the pure "symbolic" order, but he can appeal to the abstractness of mathematics to protect himself from the charge of logical circularity since his purely symbolic order does not assume any extended intuitions. But there is a price, for neither does he explain *why* we choose to represent abstract combinatorial associative–dissociative processes as extended, or what an extended representation of these processes is as opposed to any other representation, and the mystery of our representation of the extension of the physical world is still untouched.

It occurs to me now that there is a better way to think of these constructions and that is to think of the extension of the world as "co-emerging" with the extension-tracing processes that delineate it, not before (like a Kantian drafting board), not after (like a Leibnizian or Grassmannian construction of extension from scratch). The way I think of it now, perhaps space, like time, is traced out by the processes that *create* space as they co-emerge together. So when you walk around a room, taking in the various views of the room and linking them up in time order, in memory, the spatio-temporal room literally was not there before you *created* it by walking around and taking it in by stages that you later linked together and called the space of the room. There was nothing like a prior spatial room before there were perspectival space-tracing processes, there were just instantaneous moments and events, and even these moments could have been further decomposed into the instantaneous point-events and their unextended perspectival relations. The basic combinatorial laws which underlie extension, and which are given by the associative–dissociative alternations of the tracing processes, are, I believe, very fundamental physical laws which explain the origin of space-time representation, and everything that happens in them. These laws and relations can be thought of as co-emergent with space and time, not prior to them, not after, and so they can even be given a space and time formulation as a tracing process for space, time, and physical systems *in* space and time, without begging the question, even though these laws are also potentially capable of explaining what space and time *are*. This "co-emergent" approach now appears to me to be correct, and I wonder if perhaps it is what thinkers like Leibniz or Herbart or even Kant may have had in mind all along (for example, when Leibniz famously insists that space and time are *phenomena bene fundata*). Best of all, this view does not bar any fundamental inquiry into the combinatorial laws behind space and time representation, and it does not require an appeal to purely abstract symbolic relations simply to avoid the charge of circularity.

I might also point out that there are two ways to look at the extensions generated à la Grassmann, as he himself insists. We can represent the extensions as serially traced processes or we can represent the extensions as finished geometric objects or "systems" of events, like a process of change whose stages are given all at once, a length, area, or a volume, instead of traced out in a series of stages in which only the present stage is real at any one instant or slice through the object. What determines which representation of the generated extensions is the correct one? Or are they both correct, as Grassmann believed?

We might address this question by reconsidering Abbott's classic *Flatland* (1884/1992), in which extension is itself a real but perspectival property of our world. In the various Flatland universes (the three-dimensional space which is only hypothesized by the Flatlanders, the two-dimensional plane of Flatland itself, and Line- and even Pointland) the perspectival representation of the extension of objects depends in an important way on the level of extension of the observer's perspective as much as the systems of the objects themselves, or what we might call the observer's viewing pane. A hollow sphere can be seen all at once by the residents of three-dimensional space, but the residents of Flatland have to take in the sphere progressively as a series of stages in time, in which each stage of the sphere is presented sequentially, from a tangent point as the sphere first touches the viewing pane of the Flatlanders, to a series of circles growing in diameter to a maximum and then shrinking again to a point. The instantaneous time-slice is a sphere for our three-space, a point or circle for Flatland, and a line for Lineland.

It was once objected to Abbott that no one would ever observe anything in Flatland because the slices have no thickness. Abbott responded that what actually counts in spatial representation is the *relative* level of the Flatlanders, and their viewing pane or sense organs. So if their perspectival viewing pane is just as "flat" as the objects they view, there is no problem of representation, and flat objects will be seen all at once as a complete object by a flat observer, as three-dimensional beings can see three-dimensional objects; in fact this is what actually *explains* spatial representation. In Kantian terms, the level of extension an observer experiences is another subject–object perspectival relation. The level of extension of objects depends upon the extension of the viewing pane as much as the nature or complexity of the objects being viewed. So for example, if we had actual experimental access to higher-dimensional observables, we could measure higher-dimensional properties of physical systems directly, such as the superpositions of three-dimensional physical systems.

For an immanent observer within the manifold he is observing, taking in objects-to-a-subject within the space, at a lower level of extension than the objects he observes, a serially generated series of stages is necessarily what he observes in that perspective. If we are three-dimensional observers moving through space-time along our worldlines, four-dimensional solid objects would necessarily appear as serial constructions in stages, in that perspective, not as extension given all at once. But if we wish, any extension can be viewed as a construction from more basic elements all the way down to the most fundamental point-events and their qualities, even if it *appears* as a finished extension in the perspective of a higher-dimensional observer, who is himself extended relative to the object. Thus, the level of represented extension is not a perspective- or observer-independent property and is in a sense an artifact of the subject-to-object representational capacities which underlie spatial representation, according to Kant and Abbott. What *is* real are the underlying combinatorial associative–dissociative processes and laws by which extensions are actually *generated*, not the way in which we represent them, or even whether we represent these processes as serial or instantaneous. They are both phenomena, in the Greek sense of that which appears a certain way, to a certain observer, but they are also *phenomena bene fundata* in the sense that the property of extension is well founded on underlying natural laws. This, then, seems to me to be the right answer to the puzzle of extended representation raised at the beginning of this chapter: whether we regard an extension-generating process as serial or all given at once is a real but perspective-dependent fact—spatial representation is indeed just a species of intellectual subject–object representation as Kant insisted—but the extension-generating processes are real physical processes co-emergent with space and time, no matter at which level they are viewed. An actual explanation of physical space and time might therefore be possible through an investigation of those co-emergent laws and processes which we call a spatio-temporal representation of events.

Appendix: An outline of realistic empiricism

Now that the reader is familiar with the central ideas of the book and their development, in this outline I will bring together and recapitulate the various claims of realistic empiricism concisely.

Events, powers, manifestations, causal relations

A. What exist are individual events.
B. Whatever exists does so through the manifestation of power that makes it exist.
C. Individual events are a product of a power and a circumstance of manifestation. Individual events are the concrete manifestations of powers under specific circumstances of manifestation. Manifestation events include cases in which powers modify or even prevent each other's manifestations.
D. Powers are token-identical to their individual manifestations in events and identical *qua* powers *across* different token manifestations. Individual manifestations are not identical to each other *across* other token manifestations in events.
E. The concrete manifestations of power in an individual event are the event's individual qualities.
F. Events are identified one by one through their individual qualities, expressed from all possible causal perspectives on the event.
G. Events affect other events through individual causal relations, one particular event to another particular event. Particular functional relations between events are not arbitrary but are as real as the particular events they relate and are grounded in their qualities.

Point-events in perspective space

A. Each event's quality can be represented as a "potential jump" from a zero point-event (o) to the event itself (P). Any point can be chosen as the zero point (o) from which to represent the qualities of all the other point-events. Events have different qualities from different zeroes. Each event exhibits its own signature qualities or potential jumps from all other possible zeroes.

B.   Individual events are ordered into a projective space of classification by their quality types. Two events alike for one quality type will differ for others and this will serve to differentiate them as individuals within the space of events, even if there is no metric and no parallelism of directions either.

C.   Perspectives are objectively consistent with each other. The simple potential jump from o to A can be represented objectively from B's perspective as a potential jump from o to B and from A to B, even if it did not actually occur that way.

D.   Point-events in a given perspective have a potential weight and an individual potential type and, when a metric concept is given, they sum barycentrically by the summation of their effects.

E.   Perspective space is not an extended space, but a prior qualitative space of classification of individual events and individual qualities. Only present point-events in separate individualized perspectives exist.

Extensions in affine space, objects and physical systems

A.   Extended measurable objects and systems are traced out within a space of point-events with an affine structure, which preserves the parallelism of directions or vectors across differences in position. These vector qualities are "free" and make sense across different particular events: event-independent quality types.

B.   Extensions are traced out between two point-events by multiplying them using the exterior product. The primitive associative–dissociative symbolic relation between points and their qualities, or modes of determination, traces out the extension. Present events can be interpreted as "past traces" of other present-tracing processes in another temporal perspective.

C.   Extensions, such as those representing objects' physical systems, are perspective-independent and invariant across a variety of perspectives, changes of basis, and changes of position, direction, velocity, and time.

D.   The symbolic processes resulting in an extension can be regarded as a serial process of associating a point with a subsequent point, dissociating it, and reproducing and adding the previous stages to the present stage of the process. Only the unextended tracing point exists at any one time. But the whole process can also be regarded as one simultaneously given extension, like a set of nested associated–dissociated elements given all at once. The difference depends upon the level of extension of the observer's viewing pane. The represented level of extension is an observer-dependent or perspectival fact; the construction of extension from points remains valid at any level of extension.

Mental events, macro-causal relations, and sensations

A. Individual events are neutral, neither mental nor physical. Neutral events make up "physical" systems and extensions and "mental" sensations in minds through different functional relations. The functional relations do not differ in kind either, except provisionally for purposes of investigation, and can all be classed as "natural" ("mental" and "physical") functional relations between events, or simply as "physical" in an enhanced view of the physical which includes sensations among physical events.

B. Mental events are complex, configured individual events in the brain, a physical object. Mental variations are also physical variations, not in external physical objects but in the behavior of internal energies in the brain, part of the physical universe.

C. Powers (such as internal energies in neurons) manifest in individualized events a, b, c . . . (like electrical energy siphoned off into 1,000 electrodes in individual cells) but collectively in the complex individual macro-event M (a sensation of a patch of color or sound) when these powers are configured in certain ways to elicit, impede, or constrain each other. These are different non-identical empirical manifestations of the same powers: denoted by different terms "a" "b" "c" for the individualized manifestations of powers and "M" for the collective manifestations of those same powers acting in a certain configuration.

D. Configurations determine the conditions under which powers can manifest and thus play a decisive macro-causal role in determining how those powers collectively manifest when in configured circumstances. Determining manifestation conditions means playing a causal role in determining events. Configurations block, elicit, and constrain the manifestations of individual powers configured in them. Configurations also block, elicit, and constrain one another's occurrences as macro-events. Configuration-level events participate in a network of macro-causal relations with each other.

E.   Against epiphenomenalism, sensations are configuration-level macro-events. Sensation qualities are manifested powers of configuration-level events in a network of macro-causal relations. Sensations always occur in structured quality spaces. Quality spaces are macro-causal networks among configuration-level events.

F.   Against panpsychism, higher-order qualities are a separate empirical manifestation which *cannot* be deduced a priori from the empirical manifestations of lower-order qualities by a line of descent or a priori compositional principles. Sensation qualities are manifested first at the macro-causal level and in that network of macro-causal relations and not below in proto-sensations in matter.

G.   A posteriori physicalism: macro-causal powers are identical to the lower-order individualized powers that make them up in a configuration, but the micro- and macro-manifestation events are non-identical, not co-occurrent, and not related conceptually.

Epistemology and empiricist "umbrella theory" schema

A.   Epistemically, objects are prior to elements.

B.   In judgments of perception, we are in direct contact with the proper parts of external objects arranged in causal perspectives around the observer.

C.   Intellectual judgments of perception presume, for their conditions of assertibility, a system of external objects arranged in causal perspectives that radiate from the subject's immanent point of view situated *within* the perspectival system. Judgments of perception make implicit intellectual reference to future causal links and to causally linked perspectives not even possibly given to the observer in an egocentric skeptical scenario.

D.   Empirical realism: extended spatio-temporal objects in causal relations to the observer are not the *only* possibility for a perspectival system of objects or events. These objects and systems can also be reconstructed as abstract tables, matrices, and in a variety of functional forms.

E.   Scientific theories can be stripped of their visual content and mechanisms, and reduced to elementary neutral events and a variety of phenomenological laws and functions as above. This schema of elements and functions is not a scientific theory but an umbrella "theory schema" for the construction of empiricist theories in science. The realistic empiricist theory schema is an engine of analysis for eliminating inessential models and visualizations from science. We begin with the observable evidence, guess or fill in missing individual events continuous with these, and complete a perspectival system of objects around us by filling in partially observed functions. We do not, however, posit that this is the *unique* perspectival system or set of objects, nor do we posit mechanisms for every natural regularity, nor do we insist on a psychologically visualizable model of natural processes or events behind observation. Theory schemata (like Darwin's original evolutionary framework) are evaluated by their ability to predict specific theories just as theories predict data, but at a higher level of generality such as theory design.

F. Psychological theories are constructed by identifying certain kinds of complex events in the brain that are physiologically caused, and functionally relating them together by minimalistic psychological regularities or laws, such as those of association, memory, reflex, or automatic mechanisms in perception. The ego is considered the sum, or composition, or fusion, of these functional sub-systems, with no underlying "unity of consciousness."

References

Abbott, Edwin 1884/1992. *Flatland: A Romance of Many Dimensions*. New York: Dover.

Banks, Erik C. 2001. "Ernst Mach and the Episode of the Monocular Depth Sensations," *Journal of the History of the Behavioral Sciences* 37 (4) pp. 327–348.

2002. "Ernst Mach's 'New Theory of Matter' and his Definition of Mass," *Studies in History and Philosophy of Modern Physics* 33 (4) pp. 605–635.

2003. *Ernst Mach's World Elements: A Study in Natural Philosophy*. Dordrecht: Kluwer Academic Publishers.

2004. "Philosophical Roots of Ernst Mach's Economy of Thought," *Synthese* 139 (1) pp. 22–53.

2005. "Kant, Herbart and Riemann," *Kant-Studien* 96 (2) pp. 208–234.

2008. "The Problem of Extension in Natural Philosophy," *Philosophia Naturalis* 45 (2) pp. 211–235.

2010. "Neutral Monism Reconsidered," *Philosophical Psychology* 23 (2) pp. 173–187.

2012. "Sympathy for the Devil: Reconsidering Ernst Mach's Empiricism," *Metascience* 21 pp. 321–330.

2013a. "Metaphysics for Positivists: Mach versus the Vienna Circle," *Discipline Filosofiche* 23 (1) pp. 57–77.

2013b. "William James' Direct Realism: A Reconstruction," *History of Philosophy Quarterly* 30 (3) pp. 271–91.

2013c. "Extension and Measurement: A Constructivist Program from Leibniz to Grassmann," *Studies in History and Philosophy of Science A* 44 pp. 20–31.

Barbour, Julian 2000. *The End of Time* Oxford University Press.

Barbour, Julian and Pfister, Herbert (eds.) 1995. *Mach's Principle* Boston: Birkhauser.

Bell, John S. 1987. *Speakable and Unspeakable in Quantum Mechanics* Cambridge University Press.

Bird, Alexander 2007. "The Regress of Pure Powers?" *Philosophical Quarterly* 57 (22) pp. 513–534.

Blackmore, John 1973. *Ernst Mach: His Life, Work and Influence*, Berkeley: University of California Press.

(ed.) 1992. *Ernst Mach, A Deeper Look: Documents and New Perspectives* Dordrecht: Kluwer Academic Publishers.

2006. "Review of Erik Banks *Ernst Mach's World Elements*," *Isis* 97 (1) pp. 163–164.

Boring, Edwin 1942. *Sensations and Perception in the History of Experimental Psychology* New York: Appleton Century Crofts.

1950. *A History of Experimental Psychology* New York: Prentice Hall.

Boscovich, Roger 1763/1966. *A Theory of Natural Philosophy* Cambridge, MA: MIT Press.

Bostock, David 2012. *Russell's Logical Atomism* Oxford University Press.

Brewer, Bill 2002. *Perception and Reason* Oxford University Press.

Browne, John 2009. *Grassmann Algebra* Melbourne: Quantica Publishing.

Camillieri, Kristian 2009. *Heisenberg and the Interpretation of Quantum Mechanics* Cambridge University Press.

Carlson, Thomas 1950. "Empiricism, Semantics and Ontology," *Revue Internationale de Philosophie* 4 pp. 20–40.

1997. "James and the Kantian Tradition," in Putnam (ed.) pp. 363–383.

Carnap, Rudolf *et al.* 1929. *Wissenschaftliche Weltauffassung: Der Wiener Kreis* Vienna: Artur Wolf Verlag.

Chalmers, David 1996. *The Conscious Mind* Oxford University Press.

2002. "Consciousness and its Place in Nature," in D. Chalmers (ed.) *Philosophy of Mind: Classical and Contemporary Readings* Oxford University Press pp. 247–272.

2010. *The Character of Consciousness* Oxford University Press.

Cooper, Wesley 2002. *The Unity of William James' Thought* Nashville, TN: Vanderbilt University Press.

Crowe, Michael 1967. *A History of Vector Analysis* University of Notre Dame Press.

Deltete, Robert 1999. "Helm and Boltzmann: Energetics and the Lübeck Naturforscherversammlung," *Synthese* 119 (1–2) pp. 45–68.

Demopoulos, William and Friedman, Michael 1989. "The Concept of Structure in the Analysis of Matter," in *Rereading Russell: Minnesota Studies in the Philosophy of Science* 12 University of Minnesota Press pp. 183–199.

De Risi, Vincenzo 2007. *Geometry and Monadology* Basel: Birkhäuser.

Doran, Chris and Lasenby, Anthony. 2007. *Geometric Algebra for Physicists* Cambridge University Press.

Dretske, Fred 1993. "Mental Events as Structuring Causes of Behavior," in John Heil and Alfred R. Mele (eds.) *Mental Causation* Oxford University Press pp. 121–136.

Eames, Elizabeth 1967. "The Consistency of Russell's Realism," *Philosophy and Phenomenological Research* 27 (4) pp. 502–511.

Earman, John 1972. "Notes on the Causal Theory of Time," *Synthese* 24 (1–2) pp. 74–86.

Einstein, Albert 1949. "Autobiographisches," in P. Schilpp (ed.) *Albert Einstein: Philosopher Scientist* Evanston, IL: Open Court pp. 1–96.

Feigl, Herbert 1958. *The "Mental" and the "Physical"* University of Minnesota Press.

Feyerabend, Paul 1970. "Philosophy of Science: a Subject with a Great Past," *Minnesota Studies in the Philosophy of Science* 5, University of Minnesota Press.

  1984. "Mach's Theory of Research and its Relation to Einstein," *Studies in History and Philosophy of Science* 15 (1) pp. 1–22.

Finkelstein, David 1969. "Space Time Code," *Physical Review* 184 (5) pp. 1261–1271.

Friedman, Michael 1983. *Foundations of Space-Time Theories* Princeton University Press.

  1999. *Reconsidering Logical Positivism* Cambridge University Press.

  2012. "Kant on Geometry and Spatial Intuition," *Synthese* 186 (1) pp. 231–255.

Gale, Richard 1999. *The Divided Self of William James* Cambridge University Press.

Gillett, Carl and Loewer, Barry (eds.) 2001. *Physicalism and its Discontents* Cambridge University Press.

Grassmann, Hermann 1844/1995. *A New Branch of Mathematics: the Ausdehnungslehre* (Lloyd Kannenberg, trans.) Lasalle, IL: Open Court.

Grayling, A. C. 2003. "Russell, Experience, and the Roots of Science," in Nicholas Griffin (ed.) *The Cambridge Companion to Russell* Cambridge University Press pp. 449–474.

Gregory, Richard L. 1994. *Even Odder Perceptions* New York: Routledge.

Hagar, Amit and Hemmo, Meir 2013. "The Primacy of Geometry," *Studies in History and Philosophy of Modern Physics* 44 (3) pp. 357–364.

Haller, Rudolf and Stadtler, Friedrich 1988. *Ernst Mach: Werk und Wirkung* Vienna: Hölder Pichler Tempsky.

Harman, Gilbert 1990. "The Intrinsic Quality of Experience," *Philosophical Perspectives* 4 pp. 31–52.

Hatfield, Gary 1991. *The Natural and the Normative: Theories of Spatial Perception from Kant to Helmholtz* Cambridge, MA: MIT Press.

  2002. "Sense Data and the Philosophy of Mind: Russell, James and Mach," *Principia* 6 (2) pp. 203–30.

Heidelberger, Michael 2004. *Nature from Within: Gustav Theodor Fechner and His Psychophysical Worldview* Pittsburgh University Press.

Heil, John 2003. *From an Ontological Point of View* Oxford University Press.

  2005. "Dispositions," *Synthese* 144 pp. 343–356.

Heisenberg, Werner 1930. *The Physical Principles of the Quantum Theory* University of Chicago Press.

  1986. *Quantentheorie und Philosophie: Vorlesungen und Aufsätze* Stuttgart: Reclam.

Herbart, J. F. 1964. *Sämtliche Werke* (Karl Kehrbach and Otto Flügel, eds.) Aalen: Scientia Verlag.

Hestenes, David 1999. *New Foundations for Classical Mechanics* Berlin: Springer.

Hiebert, Erwin 1968. *The Conception of Thermodynamics in the Scientific Thought of Mach and Planck* Freiburg im Breisgau: Ernst Mach Institut.

1973. "Ernst Mach," *Dictionary of Scientific Biography* New York: Scribners.

Hinton, J. M. 1967. "Visual Experiences," *Mind* 76 (April) pp. 217–227.

Holton, Gerald. 1988. *Thematic Origins of Scientific Thought* Cambridge, MA: Harvard University Press.

Howard, Don 2005. "Einstein as a Philosopher of Science," *Physics Today* 58 (12) pp. 34–40.

James, Henry (ed.) 1920. *The Letters of William James* Vol. 2 Boston: Atlantic Monthly Press.

James, William 1975. *Manuscript Essays and Notes: The Works of William James* Cambridge, MA: Harvard University Press.

1977. *The Writings of William James* (J. J. McDermott, ed.) University of Chicago Press.

Jammer, Max 1963. "Review of Erwin Hiebert's *Historical Roots of the Conservation of Energy*," *British Journal for the Philosophy of Science* 14 (54) pp. 166–169.

1993. *Concepts of Space* Third Edition. New York: Dover.

1999. *Concepts of Force* New York: Dover.

Jenkin, Fleeming 1867. "Review of the Origin of Species," *The North British Review* 46 pp. 277–318.

Jordan, Thomas 2005. *Quantum Mechanics in Simple Matrix Form* New York: Dover.

Jowett, Benjamin 1871. *The Dialogues of Plato* Cambridge University Press.

Kandel, Erich, Schwartz, James, and Jessell, Thomas 2000. *Principles of Neural Science* New York: McGraw-Hill.

Kant, Immanuel 1787/1998. *Kritik der reinen Vernunft* Hamburg: Felix Meiner Verlag.

Kim, Jaegwon 1998. *Mind in a Physical World* Cambridge, MA: MIT Press.

2005. *Physicalism or Something Near Enough* Princeton University Press.

Kleinpeter, Hans 1906. "On the Monism of Professor Mach," *Monist* 16 (2) pp. 161–168.

Lamberth, David C. 1999. *William James and the Metaphysics of Experience* Cambridge University Press.

Lanczos, Cornelius 1970. *The Variational Principles of Mechanics* New York: Dover.

Landini, Gregory 2007. *Wittgenstein's Apprenticeship with Russell* Cambridge University Press.

Laudan, Larry 1981. "A Confutation of Convergent Realism," *Philosophy of Science* 48 (1) pp. 19–49.

Leibniz, G. W. 1989. *Philosophical Essays* (Roger Ariew and Daniel Garber, trans. and eds.) Indianapolis: Hackett.

Lenin, V. I. 1908/1952. *Materialism and Empirio-Criticism* Moscow: World Languages Publishing House.

Lewis, A. 1977. "H. Grassmann's Ausdehnungslehre and Schleiermacher's Dialektik," *Annals of Science* 34 pp. 103–162.

Lockwood, Michael 1981. "What Was Russell's Neutral Monism?" *Midwest Studies in Philosophy of Science* 6 pp. 145–158.

1989. *Mind, Brain and the Quantum* Oxford: Basil Blackwell.

1993. "The Grain Problem," in H. Robinson (ed.) *Objections to Physicalism* Oxford University Press pp. 271–291.

Lovejoy, Arthur O. 1930. *The Revolt Against Dualism* Chicago: Open Court.

Mach, Ernst 1872/1910. *The History and Root of the Principle of the Conservation of Energy* (P. E. B. Jourdain, trans.) Chicago: Open Court.

1883/1960. *The Science of Mechanics: Historico-Critically Presented* (Thomas McCormack, trans.) Lasalle, IL: Open Court.

1886/1959. *The Analysis of Sensations* (C. M. Williams and Sydney Waterlow, trans.) New York: Dover.

1896/1986. *Principles of the Theory of Heat* (Thomas McCormack, trans.) Dordrecht: Reidel.

1905/1976. *Knowledge and Error* (Thomas McCormack and Paul Foulkes, trans.) Dordrecht: Reidel.

1910. "Die Leitgedanken meiner naturwissenschaftlichen Erkenntnislehre und ihre Aufnahme durch die Zeitgenossen," *Scientia* 7 pp. 225–240.

Martin, C. B. 1993. "Power for Realists," in John Bacon, Keith Campbell, and Lloyd Reinhardt (eds.) *Ontology, Causality and Mind: Essays for D.M. Armstrong* Cambridge University Press pp. 175–185.

Maxwell, Grover 1978. "Rigid Designators and Mind-Brain Identity," in C. Wade Savage (ed.) *Minnesota Studies in the Philosophy of Science 9* University of Minnesota Press pp. 365–403.

McKitrick, Jennifer 2003. "The Bare Metaphysical Possibility of Bare Dispositions," *Philosophy and Phenomenological Research* 66 (2) pp. 349–369.

McLaughlin, Brian 1992. "The Rise and Fall of British Emergentism," in Ansgar Beckermann, Hans Flohr, and Jaegwon Kim (eds.) *Emergence or Reduction: Essays on the Prospects of Nonreductive Physicalism* Berlin: Walter de Gruyter pp. 49–93.

Mumford, Stephen 1998. *Dispositions* Oxford University Press.

Newman, Max 1928. "Mr. Russell's Causal Theory of Perception," *Mind* 37 pp. 137–148.

Norton, John 1995. "Mach's Principle Before Einstein," in Barbour and Pfister (eds.) pp. 9–57.

Papineau, David 1993. *Philosophical Naturalism* Oxford: Basil Blackwell.

Peirce, Charles Sanders 1868. "Questions Concerning Certain Faculties Claimed for Man," *Journal of Speculative Philosophy* 2 pp. 103–114.

Penrose, Roger 2004. *The Road to Reality* Oxford University Press.

Petsche, Hans Joachim 2011. *From Past to Future: Grassmann's Work in Context* Basel: Springer.

(ed.) 2012. "Schleiermacher, Fries, Herbart: wer beeinflusste Hermann Grassmann bei der philosophischen Anlagen seiner Ausdehnungslehre von 1844?" *Math Semesterbericht* 59 pp. 183–222.

Pojman, Paul 2000. "Ernst Mach's Biological Theory of Knowledge" Doctoral dissertation Indiana University.

Preston, John 2008. "Mach and Hertz's Mechanics," *Studies in History and Philosophy of Science A* 39 (1) pp. 91–101.

Prior, Elizabeth 1985. *Dispositions* Aberdeen University Press.

Prior, Elizabeth, Pargetter, Robert, and Jackson, Frank 1982. "Three Theses About Dispositions," *American Philosophical Quarterly* 19 pp. 251–7.

Putnam, Hilary 1981. *Reason, Truth and History* Cambridge University Press.

1990. "James' Theory of Perception," in James Conant (ed.) *Realism With a Human Face* Cambridge, MA: Harvard University Press pp. 232–251.

1994. "Sense, Nonsense and the Senses: An Inquiry into the Powers of the Human Mind," *Journal of Philosophy* 91 pp. 445–517.

Putnam, Ruth Anna (ed.) 1997. *The Cambridge Companion to William James* Cambridge University Press.

Quine, W. V. O. 1966. "Russell's Ontological Development," *Journal of Philosophy* 63 (21) pp. 657–667.

Ratliff, Floyd 1965. *Mach Bands* San Francisco: Holden Day.

Reichenbach, Hans 1927/1956. *The Direction of Time* University of Los Angeles Press.

Richardson, Alan and Thomas Uebel (eds.) 2007. *The Cambridge Companion to Logical Positivism* Cambridge University Press.

Riemann, Bernhard 1867. "Uber die Hypothesen welche der Geometrie zu Grunde liegen," in H. Weber and R. Dedekind (eds.) *Riemann's Gesammelte Werke und Wissenschaftlicher Nachlass* Leipzig: B.G. Teubner.

Robb, A. A. 1913. *A Theory of Time and Space* Cambridge University Press.

Rosenberg, Gregg 2004. *A Place for Consciousness* Oxford University Press.

Russell, Bertrand 1897/1996. *An Essay on the Foundations of Geometry* London: Routledge.

1903. *The Principles of Mathematics* Cambridge University Press.

1912. *The Problems of Philosophy* London: The Home University Library, Williams and Norgate

1913/1926. *Our Knowledge of the External World* London: George Allen and Unwin.

1914/1957. "The Relation of Sense Data to Physics," in *Mysticism and Logic* New York: Anchor Books pp. 140–173.

1914/1984. *Collected Papers of Bertrand Russell* Vol. 7 London: George Allen and Unwin.

1918/1985. *The Philosophy of Logical Atomism* LaSalle, IL: Open Court.

1919. "On Propositions: What They Are and How They Mean," *Proceedings of the Aristotelian Society*, Supplementary Volumes Vol. 2 pp. 1–43.

1921. *The Analysis of Mind* London: Routledge.

1927/1954. *The Analysis of Matter* New York: Dover.

1947/1997. "The Principle of Individuation," in *Works of Bertrand Russell* vol. 11 (John Slater and Peter Kollner, eds.) London: Routledge.

1959/1997. *My Philosophical Development* London: Routledge.

1968/1998. *Autobiography* Vol. 2 London: Routledge.

Rynasiewicz, Robert 1996. "Absolute Versus Relational Space-Time: An Outmoded Debate?" *Journal of Philosophy* 93 (6) pp. 279–306.

Schlick, Moritz 1925/1985. *Allgemeine Erkenntnislehre* (A. Blumberg, trans.) LaSalle, IL: Open Court.

Scholz, Erhard 1982. "Herbart's Influence on Bernhard Riemann," *Historia Mathematica* 9 pp. 413–440.

Schubring, Kurt (ed.) 1996. *Hermann Günther Grassmann (1809–1877): Visionary Mathematician, Scientist and Neohumanist Scholar: Papers from a Sesquicentennial Conference* Dordrecht: Kluwer Academic Publishers.

Seager, William 2006. "The 'Intrinsic Nature' Argument for Panpsychism," *Journal of Consciousness Studies* 13, *10–11* pp. 3–31.

Sellars, Wilfred 1956/1997. *Empiricism and the Philosophy of Mind* (Robert Brandom ed.) Cambridge, MA: Harvard University Press.

Smith, Barry 1994. *Austrian Philosophy: The Legacy of Franz Brentano* Chicago: Open Court.

Sprigge, Timothy 1997. "James, Aboutness and his British Critics," in Putnam (ed.) pp. 125–144.

Stace, W. T. 1946/1999. "Russell's Neutral Monism," in P. Schilpp (ed.) *The Philosophy of Bertrand Russell* LaSalle, IL: Open Court pp. 353–384.

Stadtler, Friedrich 1997/2001. *The Vienna Circle: Studies in the Origins, Development and Influence of Logical Empiricism* Vienna: Springer.

Stoljar, Daniel 2001. "Two Conceptions of the Physical," *Philosophy and Phenomenological Research* 62 (2) pp. 253–281.

2010. *Physicalism* New York: Routledge.

Stout, G. F. 1911. "The Object of Thought and Real Being," *Proceedings of the Aristotelian Society* 11 pp. 187–208.

Strawson, Galen 2006. "Realistic Monism: Why Physicalism Entails Panpsychism," *Journal of Consciousness Studies* 13 (10–11) pp. 3–31.

Stubenberg, Leopold 2010. "Neutral Monism," *Stanford Encyclopedia of Philosophy* plato.stanford.edu/entries/neutral-monism/

Suppe, Frederick (ed.) 1977. *The Structure of Scientific Theories* University of Illinois Press.

Swimmer, Alvin 1996. "The Completion of Grassmann's *Natur-Wissenschaftliche Methode*," in Schubring (ed.) pp. 265–280.

Thiele, Joachim 1978. *Wissenschaftliche Kommunikation: Die Korrespondenz Ernst Machs* Kastellaun: Henn Verlag.

Tully, Robert E. 1993. "Three Studies of Russell's Neutral Monism," *(Parts I,II) Russell* 13: 5–35; 185–202.

1999. "Russell's Neutral Monism," in Andrew Irvine (ed.) *Bertrand Russell: Critical Assessments* Vol. 3 London: Routledge pp. 209–224.

Uebel, Thomas 2007. *Empiricism at the Crossroads: The Vienna Circle's Protocol Sentence Debate* Chicago: Open Court.

Unger, Peter 1999. "The Mystery of the Physical and the Matter of Qualities: A Paper for Professor Shaffer," *Midwest Studies in Philosophy* 23 (1) pp. 75–99.

Von Békésy, Georg 1960. *Experiments in Hearing* New York: McGraw-Hill.

Weinberg, Carlton Berenda 1937. *Mach's Empirio-Pragmatism in Physical Science* New York: Albee Press.

Weyl, Hermann 1920. "Allgemeine Diskussion über die Relativitätstheorie," *Physikalische Zeitschrift* 21 pp. 666–668.

1922. "Die Relativitätstheorie auf der Naturforscherversammlung," in *Bad Nauheim Jahresbericht der Deutschen Mathematiker-Vereinigung* 31 pp. 51–63.

Winnie, John 1977. "The Causal Theory of Spacetime," in John S. Earman, Clark N. Glymour, and John J. Stachel (eds.) *Foundations of Spacetime Theories* (Minnesota Studies in the Philosophy of Science) University of Minnesota Press.

Wittgenstein, Ludwig 1995. *Ludwig Wittgenstein: Cambridge Letters* (B. McGuinness and G. H. von Wright, eds.) Oxford: Basil Blackwell.

Zaddach, Arno 1994. *Grassmanns Algebra in der Geometrie* Mannheim: Wissenschaftliche Verlag.

Index